Handbook of
Internal Combustion Engines

Handbook of Internal Combustion Engines

Felix Pierson

Oxford Book Company
Jaipur ▪ New Delhi

ISBN : 9789355241863

First published in India, 2025

Oxford Book Company
Regd. Off. "Bony Residency", Gate No. 2
Opp. Tilak Public School, Vishweswariya Nagar,
Gopalpura Road, Jaipur - 302018, India
Ph: 0141 - 2761280, Telefax: 2761381
email - oxfordbookjpr@gmail.com
website - oxfordbookcompany.com

Delhi Office: 102, 1st Floor, Satyam House,
4327/3, Ansari Road, Darya Ganj,
New Delhi - 110002, India
Ph: 011-45652440

Distributed By
KEYA INTERNATIONAL PRESS LLP
E-mail: keyainternationalpressllp@gmail.com

Printed and bound in India by Thomson Press Ltd., New Delhi

Contents

Preface

The term *internal combustion engine* usually refers to an engine in which combustion is intermittent, such as the more familiar four-stroke and two-stroke piston engines, along with variants, such as the six-stroke piston engine and the Wankel rotary engine. A second class of internal combustion engines use continuous combustion: gas turbines, jet engines and most rocket engines, each of which are internal combustion engines on the same principle as previously described. The internal combustion engine (or ICE) is quite different from external combustion engines, such as steam or Stirling engines, in which the energy is delivered to a working fluid not consisting of, mixed with, or contaminated by combustion products. Working fluids can be air, hot water, pressurized water or even liquid sodium, heated in some kind of boiler. A large number of different designs for ICEs have been developed and built, with a variety of different strengths and weaknesses. Powered by an energy-dense fuel. While there have been and still are many stationary applications, the real strength of internal combustion engines is in mobile applications and they dominate as a power supply for cars, aircraft, and boats. Internal combustion engines are most commonly used for mobile propulsion in vehicles and portable machinery. In mobile equipment, internal combustion is advantageous since it can provide high power-to-weight ratios together with excellent fuel energy density. Generally using fossil fuel (mainly petroleum), these engines have appeared in transport in almost all vehicles. Where very high power-to-weight ratios are required, internal combustion engines appear in the form of gas turbines. These applications include jet aircraft, helicopters, large ships and electric generators.

All internal combustion engines depend on the combustion of a chemical fuel, typically with oxygen from the air. The combustion process typically results in the production of a great quantity of heat, as well as the production of steam and carbon dioxide and other chemicals at very high temperature; the temperature reached is determined by the chemical make up of the fuel and oxidisers, as well as by the compression and other factors. The most common modern fuels are made up of hydrocarbons and are derived mostly from fossil fuels (petroleum). Fossil fuels include diesel fuel, gasoline and petroleum gas, and the rarer use of propane. Except for the fuel delivery components, most internal combustion engines that are designed for gasoline use can run on natural gas or liquefied petroleum gases without major modifications. Large diesels can run with air mixed with gases and a pilot diesel fuel ignition injection. Liquid and gaseous biofuels, such as ethanol and biodiesel (a form of diesel fuel that is produced from crops that yield triglycerides such as soybean oil), can also be used. Engines with appropriate modifications can also run on hydrogen gas, wood gas, or charcoal gas, as well as from so-called producer gas made from other convenient biomass. Recently, experiments have been made with using powdered solid fuels, such as the magnesium injection cycle. Internal combustion engines require ignition of the mixture, either by spark ignition (SI) or compression ignition (CI). Before the invention of reliable electrical methods, hot tube and flame methods were used. Experimental engines with laser ignition have been built. Gasoline engine ignition systems generally rely on a combination of a lead–acid battery and an induction coil to provide a high-voltage electric spark to ignite the air-fuel mix in the engine's cylinders. This battery is recharged during operation using an electricity-generating device such as an alternator or generator driven by the engine. Gasoline engines take in a mixture of air and gasoline and compress it to not more than 12.8 bar (1.28 MPa), then use a spark plug to ignite the mixture when it is compressed by the piston head in each cylinder. Diesel engines and HCCI engines, rely solely on heat and pressure created by the engine in its compression process for ignition. The compression level that occurs is usually twice or more than a gasoline engine. Diesel engines will take in air only, and shortly before peak compression, a small quantity of diesel fuel is sprayed into the cylinder via a fuel injector that allows the fuel to instantly ignite. HCCI type engines will take in both air and fuel but continue to rely on an unaided auto-combustion process, due to higher pressures and heat.

This Handbook will take the readers back to the past 20 years where automakers had significantly improved its power, its fuel efficiency, and its emissions, with more changes to come.

— *Felix Pierson*

Chapter 1

Introduction

Internal Combustion Engine

The internal combustion engine is an engine in which the combustion of a fuel (normally a fossil fuel) occurs with an oxidiser (usually air) in a combustion chamber. In an internal combustion engine, the expansion of the high-temperature and high -pressure gases produced by combustion apply direct force to some component of the engine.

This force is applied typically to pistons, turbine blades, or a nozzle. This force moves the component over a distance, transforming chemical energy into useful mechanical energy.

The term *internal combustion engine* usually refers to an engine in which combustion is intermittent, such as the more familiar four-stroke and two-stroke piston engines, along with variants, such as the six-stroke piston engine and the Wankel rotary engine. A second class of internal combustion engines use continuous combustion: gas turbines, jet engines and most rocket engines, each of which are internal combustion engines on the same principle as previously described.

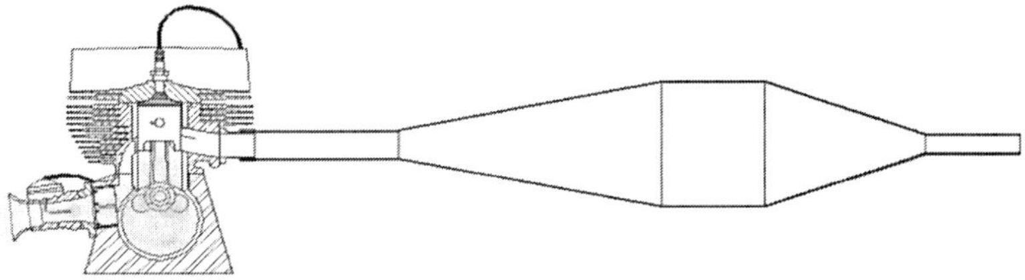

Figure 1: Animated two-stroke engine in operation

The internal combustion engine (or ICE) is quite different from external combustion engines, such as steam or Stirling engines, in which the energy is delivered to a working fluid not consisting of, mixed with, or contaminated by combustion products. Working fluids can be air, hot water, pressurised water or even liquid sodium, heated in some kind of boiler. A large number of different designs for ICEs have been developed and built, with a variety of different strengths and weaknesses. Powered by an energy-dense fuel (which is very frequently gasoline, a liquid derived from fossil fuels). While there have been and still are many stationary applications, the real strength of internal combustion engines is in mobile applications and they dominate as a power supply for cars, aircraft, and boats.

Applications

Internal combustion engines are most commonly used for mobile propulsion in vehicles and portable machinery. In mobile equipment, internal combustion is advantageous since it can provide high power-to-weight ratios together with excellent fuel energy density.

Generally using fossil fuel (mainly petroleum), these engines have appeared in transport in almost all vehicles (automobiles, trucks, motorcycles, boats, and in a wide variety of aircraft and locomotives). Where very high power-to-weight ratios are required, internal combustion engines appear in the form of gas turbines. These applications include jet aircraft, helicopters, large ships and electric generators.

History of the Internal Combustion Engine

Although various forms of internal combustion engines were developed before the 19th century, their use was hindered until the commercial drilling and production of petroleum began in the mid-1850s. By the late 19th century, engineering advances led to their widespread adoption in a variety of applications.

Timeline of Development

Various scientists and engineers contributed to the development of internal combustion engines:

Prior to Modern Era

- 13th century: The rocket engine, an internal-combustion engine, was developed by the Chinese, Mongols and Arabs.

1600 to 1860

- 17th century: Christiaan Huygens designs gunpowder to drive water pumps, to supply 3000 cubic metres of water/day for the Versailles palace gardens, essentially creating the first idea of a rudimentary internal combustion piston engine.

- 1780s: Alessandro Volta built a toy electric pistol in which an electric spark exploded a mixture of air and hydrogen, firing a cork from the end of the gun.
- 1791: John Barber receives British patent #1833 for *A Method for Rising Inflammable Air for the Purposes of Producing Motion and Facilitating Metallurgical Operations*. In it he describes a turbine.
- 1794: Robert Steele built a compressionless engine whose principle of operation would dominate for nearly a century.
- 1798: Tippu Sultan, the ruler of the city-state of Mysore in India, uses the first iron rockets against the British Army.
- 1807: Nicephore Niepce installed his 'moss, coal-dust and resin' fuelled Pyreolophore internal combustion engine in a boat and powered up the river Saône in France. A patent was subsequently granted by Emperor Napoleon Bonaparte on 20 July 1807.
- 1807: Swiss engineer Francois Isaac de Rivaz built an internal combustion engine powered by a hydrogen and oxygen mixture, and ignited by electric spark.
- 1823: Samuel Brown patented the first internal combustion engine to be applied industrially. It was compressionless and based on what Hardenberg calls the "Leonardo cycle," which, as the name implies, was already out of date at that time.
- 1824: French physicist Sadi Carnot established the thermodynamic theory of idealised heat engines. This scientifically established the need for compression to increase the difference between the upper and lower working temperatures.
- 1826 April 1: American Samuel Morey received a patent for a compressionless "Gas or Vapour Engine."
- 1833: Lemuel Wellman Wright, UK patent 6525, table-type gas engine. Double acting gas engine, first record of water jacketed cylinder.
- 1838: A patent was granted to William Barnett (English). According to Dugald Clerk, this was the first recorded use of in-cylinder compression.
- 1854-57: Eugenio Barsanti & Felice Matteucci invented an engine that was possibly the first 4-cycle engine, but the patent was lost.

- 1856: in Florence at *Fonderia del Pignone* (now Nuovo Pignone, later a subsidiary of General Electric), Pietro Benini realised a working prototype of the Italian engine supplying 5 HP. In subsequent years he developed more powerful engines—with one or two pistons—which served as steady power sources, replacing steam engines.
- 1857: Eugenio Barsanti & Felice Matteucci describe the principles of the free piston engine where the vacuum after the explosion allows atmospheric pressure to deliver the power stroke (British patent No 1625). Otto and Langen were the first to make a marketable engine based on this concept 10 years later.

1860-1910

- 1860: Belgian Jean Joseph Etienne Lenoir (1822–1900) produced a gas-fired internal combustion engine similar in appearance to a horizontal double-acting steam engine, with cylinders, pistons, connecting rods, and flywheel in which the gas essentially took the place of the steam. This was the first internal combustion engine to be produced in numbers.
- 1861 The earliest confirmed patent of the 4-cycle engine, by Alphonse Beau de Rochas. A year earlier, Christian Reithmann made an engine which may have been the same, but it's unknown since his patent wasn't clear on this point.
- 1862: German inventor Nikolaus Otto was the first to build and sell the engine. He designed an indirect-acting free-piston compressionless engine whose greater efficiency won the support of Eugen Langen and then most of the market, which at that time was mostly for small stationary engines fuelled by lighting gas.
- 1865: Pierre Hugon started production of the Hugon engine, similar to the Lenoir engine, but with better economy, and more reliable flame ignition.
- 1867: Otto and Langen introduced their free piston engine at the Paris Exhibition. It had less than half the gas consumption of the Lenoir or Hugon engines.
- 1870: In Vienna, Siegfried Marcus put the first mobile gasoline engine on a handcart.
- 1872: In America George Brayton invented Brayton's Ready Motor and went into commercial production, this used constant

pressure combustion, and was the first commercial liquid fuelled internal combustion engine.

- 1876: Nikolaus Otto, working with Gottlieb Daimler and Wilhelm Maybach, started the genesis of the four-cycle engine. The German courts, however, did not hold his patent to cover all in-cylinder compression engines or even the four-stroke cycle, and after this decision, in-cylinder compression became universal.
- 1878: Dugald Clerk designed the first two-stroke engine with in-cylinder compression. He patented it in England in 1881.
- 1879: Karl Benz, working independently, was granted a patent for his internal combustion engine, a reliable two-stroke gas engine, based on the same technology as De Rochas's design of the four-stroke engine. Later, Benz designed and built his own four-stroke engine that was used in his automobiles, which were developed in 1885, patented in 1886, and became the first automobiles in production.
- 1882: James Atkinson invented the Atkinson cycle engine. Atkinson's engine had one power phase per revolution together with different intake and expansion volumes, potentially making it more efficient than the Otto cycle, but certainly avoiding Otto's patent.
- 1884: British engineer Edward Butler constructed the first petrol (gasoline) internal combustion engine. Butler invented the spark plug, magneto, coil ignition and spray jet carburettor, and was the first to use the word petrol.
- 1885: German engineer Gottlieb Daimler received a German patent for a supercharger
- 1891: Herbert Akroyd Stuart built his oil engine, leasing rights to Hornsby of England to build them. They built the first cold-start compression-ignition engines. In 1892, they installed the first ones in a water pumping station. In the same year, an experimental higher-pressure version produced self-sustaining ignition through compression alone.
- 1892: Dr. Rudolf Diesel developed his Carnot heat engine type motor.
- 1887: Gustaf de Laval introduces the de Laval nozzle

- 1893 February 23: Rudolf Diesel received a patent for his compression ignition (diesel) engine.
- 1896: Karl Benz invented the boxer engine, also known as the horizontally opposed engine, or the flat engine, in which the corresponding pistons reach top dead centre at the same time, thus balancing each other in momentum.
- 1900: Rudolf Diesel demonstrated the diesel engine in the 1900 *Exposition Universelle* (World's Fair) using peanut oil fuel.
- 1900: Wilhelm Maybach designed an engine built at Daimler Motoren Gesellschaft—following the specifications of Emil Jellinek—who required the engine to be named *Daimler-Mercedes* after his daughter. In 1902 automobiles with that engine were put into production by DMG.
- 1903 - Konstantin Tsiolkovsky begins a series of theoretical papers discussing the use of rocketry to reach outer space. A major point in his work is liquid fuelled rockets.
- 1903: Ægidius Elling builds a gas turbine using a centrifugal compressor which runs under its own power. By most definitions, this is the first working gas turbine.
- 1905 Alfred Buchi patents the turbocharger and starts producing the first examples.
- 1903-1906: The team of Armengaud and Lemale in France build a complete gas turbine engine. It uses three separate compressors driven by a single turbine. Limits on the turbine temperatures allow for only a 3:1 compression ratio, and the turbine is not based on a Parsons-like "fan", but a Pelton wheel-like arrangement. The engine is so inefficient, at about 3% thermal efficiency, that the work is abandoned.
- 1908: New Zealand inventor Ernest Godward started a motorcycle business in Invercargill and fitted the imported bikes with his own invention – a petrol economiser. His economisers worked as well in cars as they did in motorcycles.
- 1908: Hans Holzwarth starts work on extensive research on an "explosive cycle" gas turbine, based on the Otto cycle. This design burns fuel at a constant volume and is somewhat more efficient. By 1927, when the work ended, he has reached about 13% thermal efficiency.
- 1908: Rene Lorin patents a design for the ramjet engine.

1910-1960

- 1916: Auguste Rateau suggests using exhaust-powered compressors to improve high-altitude performance, the first example of the turbocharger.
- 1920: William Joseph Stern reports to the Royal Air Force that there is no future for the turbine engine in aircraft. He bases his argument on the extremely low efficiency of existing compressor designs. Due to Stern's eminence, his paper is so convincing there is little official interest in gas turbine engines anywhere, although this does not last long.
- 1921: Maxime Guillaume patents the axial-flow gas turbine engine. It uses multiple stages in both the compressor and turbine, combined with a single very large combustion chamber.
- 1923: Edgar Buckingham at the United States National Bureau of Standards publishes a report on jets, coming to the same conclusion as W.J. Stern, that the turbine engine is not efficient enough. In particular he notes that a jet would use five times as much fuel as a piston engine.
- 1925: The Hesselman engine is introduced by Swedish engineer Jonas Hesselman represented the first use of direct gasoline injection on a spark-ignition engine.
- 1925: Wilhelm Pape patents a constant-volume engine design.
- 1926: Alan Arnold Griffith publishes his groundbreaking paper *Aerodynamic Theory of Turbine Design*, changing the low confidence in jet engines. In it he demonstrates that existing compressors are "flying stalled", and that major improvements can be made by redesigning the blades from a flat profile into an airfoil, going on to mathematically demonstrate that a practical engine is definitely possible and showing how to build a turboprop.
- 1926 - Robert Goddard launches the first liquid fuelled rocket
- 1927: Aurel Stodola publishes his "Steam and Gas Turbines" - basic reference for jet propulsion engineers in the USA.
- 1927: A testbed single-shaft turbo-compressor based on Griffith's blade design is tested at the Royal Aircraft Establishment.
- 1929: Frank Whittle's thesis on jet engines is published
- 1930: Schmidt patents a pulse-jet engine in Germany.

- 1936: French engineer Rene Leduc, having independently re-discovered Rene Lorin's design, successfully demonstrates the world's first operating ramjet.
- 1937: The first successful run of Sir Frank Whittle's gas turbine for jet propulsion.
- March, 1937: The Heinkel HeS 1 experimental hydrogen fuelled centrifugal jet engine is tested at Hirth.
- 27 August, 1939: The Heinkel He 178 V1 pioneer turbojet aircraft prototype makes its first flight, powered by an He S 3 engine.
- July 18, 1942: The Messerschmitt Me 262 first jet engine flight
- 1946: Sam Baylin develops the Baylin Engine a three cycle internal combustion engine with rotary pistons. A crude but complex example of the future Wankel engine.
- 1954: Felix Wankel's first working prototype DKM 54 of the Wankel engine

1960 to Present

- 1986 Benz Gmbh files for patent protection for a form of Scotch yoke engine and begins development of same. Development subsequently abandoned.
- 1999: Brothers, Michael and Peter Raffaele file patent application seeking protection for new form of Scotch yoke engine known as the Slider Engine.
- 2004 Hyper-X first scramjet to maintain altitude
- 2004 Toyota Motor Corp files for patent protection for new form of Scotch yoke engine.

Engine Starting

Early internal combustion engines were started by hand cranking. Various types of starter motor were later developed. These included:

- An auxiliary petrol engine for starting a larger petrol or diesel engine. The Hucks starter is an example
- Cartridge starters, such as the Coffman engine starter, which used a device like a blank shotgun cartridge. These were popular for aircraft engines
- Pneumatic starters
- Hydraulic starters
- Electric starters.

Electric starters are now almost universal for small and medium-sized engines, while compressed-air starting is used for large engines.

Modern vs. Historical Piston Engines

The first piston engines did not have compression, but ran on an air-fuel mixture sucked or blown in during the first part of the intake stroke. The most significant distinction between modern internal combustion engines and the early designs is the use of compression and, in particular, in-cylinder compression.

Nomenclature : At one time, the word, "Engine" (from Latin, via Old French, *ingenium*, "ability") meant any piece of machinery—a sense that persists in expressions such as *siege engine*. A "motor" (from Latin *motor*, "mover") is any machine that produces mechanical power. Traditionally, electric motors are not referred to as "Engines"; however, combustion engines are often referred to as "motors." (An *electric engine* refers to a locomotive operated by electricity.)

Types of Internal Combustion Engine

Engines can be classified in many different ways: By the engine cycle used, the layout of the engine, source of energy, the use of the engine, or by the cooling system employed.

Engine Configurations: Internal combustion engines can be classified by their configuration. Common layouts of engines are:

Reciprocating Engine

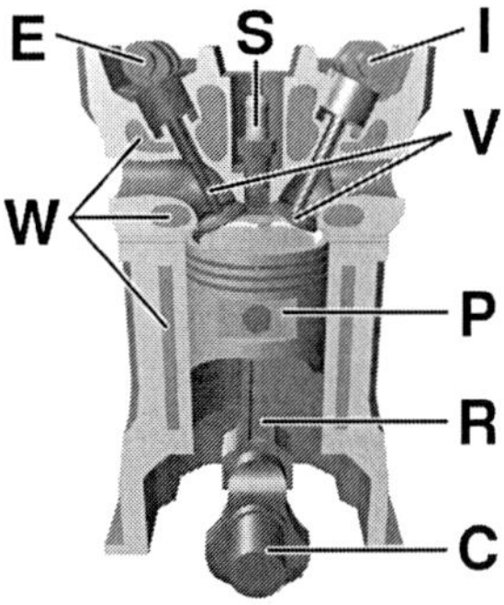

Figure 2: Internal combustion piston engine

Components of a typical, four stroke cycle, internal combustion piston engine.

E - Exhaust camshaft	I - Intake camshaft
S - Spark plug	V - Valves
P - Piston	R - Connecting rod
C - Crankshaft	W - Water jacket for coolant flow.

A reciprocating engine, also often known as a piston engine, is a heat engine that uses one or more reciprocating pistons to convert pressure into a rotating motion. This article describes the common features of all types. The main types are: the internal combustion engine, used extensively in motor vehicles; the steam engine, the mainstay of the Industrial Revolution; and the niche application Stirling engine.

Common Features in all Types

There may be one or more pistons. Each piston is inside a cylinder, into which a gas is introduced, either already hot and under pressure (steam engine), or heated inside the cylinder either by ignition of a fuel air mixture (internal combustion engine) or by contact with a hot heat exchanger in the cylinder (Stirling engine). The hot gases expand, pushing the piston to the bottom of the cylinder. The piston is returned to the cylinder top (Top Dead Centre) either by a flywheel or the power from other pistons connected to the same shaft. In most types the expanded or "exhausted" gases are removed from the cylinder by this stroke. The exception is the Stirling engine, which repeatedly heats and cools the same sealed quantity of gas. In some designs the piston may be powered in both directions in the cylinder in which case it is said to be double acting.

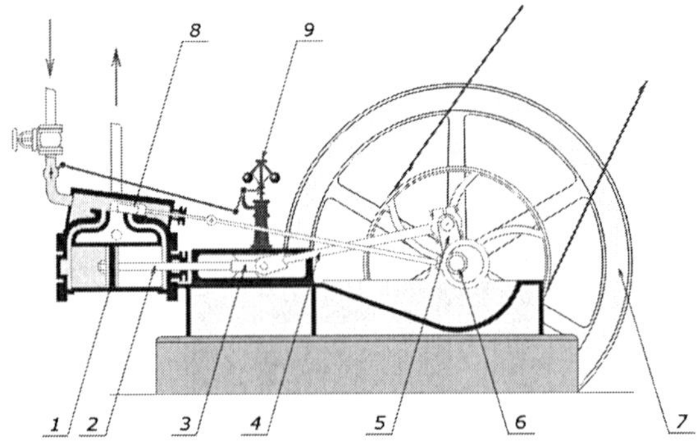

Figure 3: Steam piston engine

A labelled schematic diagram of a typical single-cylinder, simple expansion, double-acting high pressure steam engine. Power takeoff from the engine is by way of a belt.

1 – Piston
2 – Piston rod
3 – Crosshead bearing
4 – Connecting rod
5 – Crank
6 – Eccentric valve motion
7 – Flywheel
8 – Sliding valve
9 – Centrifugal governor

In all types, the linear movement of the piston is converted to a rotating movement via a connecting rod and a crankshaft or by a swashplate. A flywheel is often used to ensure smooth rotation. The more cylinders a reciprocating engine has, generally, the more vibration-free (smoothly) it can operate. The power of a reciprocating engine is proportional to the volume of the combined pistons' displacement.

A seal needs to be made between the sliding piston and the walls of the cylinder so that the high pressure gas above the piston does not leak past it and reduce the efficiency of the engine. This seal is provided by one or more piston rings. These are rings made of a hard metal which are sprung into a circular groove in the piston head. The rings fit tightly in the groove and press against the cyinder wall to form a seal.

It is common for such engines to be classified by the number and alignment of cylinders and the total volume of displacement of gas by the pistons moving in the cylinders usually measured in cubic centimetres (cm^3 or cc) or litres (l) or (L) (US: litre). For example for internal combustion engines, single and two-cylinder designs are common in smaller vehicles such as motorcycles, while automobiles typically have between four and eight, and locomotives, and ships may have a dozen cylinders or more. Cylinder capacities may range from 10 cm^3 or less in model engines up to several thousand cubic centimetres in ships' engines.

The compression ratio is a measure of the performance in an internal-combustion engine or a Stirling Engine. It is the ratio between the volume of the cylinder, when the piston is at the bottom of its stroke, and the volume when the piston is at the top of its stroke.

The bore/stroke ratio is the ratio of the diametre of the piston, or "bore", to the length of travel within the cylinder, or "stroke". If this is around 1 the engine is said to be "square", if it is greater than 1, i.e. the bore is larger than the stroke, it is "oversquare". If it is less than 1, i.e. the stroke is larger than the bore, it is "undersquare".

Cylinders may be aligned in line, in a V configuration, horizontally opposite each other, or radially around the crankshaft.

Opposed-piston engines put two pistons working at opposite ends of the same cylinder and this has been extended into triangular arrangements such as the Napier Deltic. Some designs have set the cylinders in motion around the shaft.

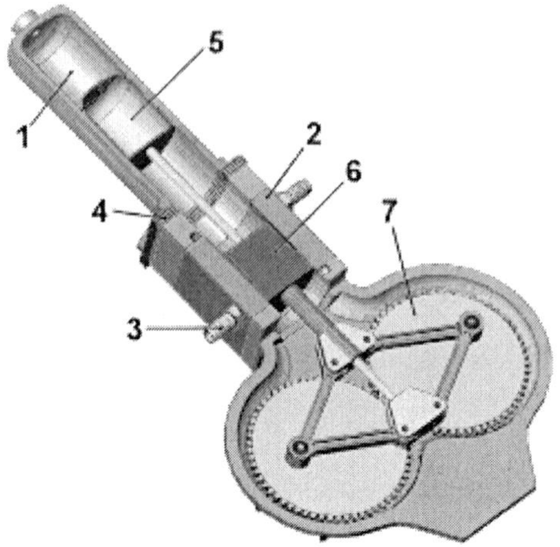

Figure 4: Stirling piston engine

Rhombic Drive – Beta Stirling Engine Design showing the second displacer piston (green) within the cylinder which shunts the working gas between the hot and cold ends, but produces no power itself.

Pink – Hot cylinder wall
Dark grey – Cold cylinder wall
Green – Displacer piston
Dark blue – Power piston
Light blue – Flywheels

In steam engines and internal combustion engines, valves are required to allow the entry and exit of gasses at the correct time in the piston's cycle. These are worked by cams or cranks driven by the shaft of the engine. Early designs used the D slide valve but this has been largely superseded by Piston valve or Poppet valve designs. In steam engines the point in the piston cycle at which the steam inlet valve closes is called the cutoff and this can often be controlled to adjust the torque supplied by the engine.

Internal combustion engines operate through a sequence of strokes which admit and remove gases to and from the cylinder. These operations are repeated cyclically and an engine is said to be 2-stroke, 4-stroke or 6-stroke depending on the number of strokes it takes to complete a cycle.

In some steam engines, the cylinders may be of varying size with the smallest bore cylinder working the highest pressure steam. This is then fed through one or more, increasingly larger bore cylinders successively, to extract power from the steam at increasingly lower pressures. These engines are called Compound engines.

History

An early known example of rotary to reciprocating motion can be found in a number of Roman saw mills (dating to the 3rd to 6th century AD) in which a crank and connecting rod mechanism converted the rotary motion of the waterwheel into the linear movement of the saw blades. The reciprocating engine developed in Europe during the 18th century, first as the atmospheric engine then later as the steam engine. These were followed by the Stirling engine and internal combustion engine in the 19th century. Today the most common form of reciprocating engine is the internal combustion engine running on the combustion of petrol, diesel, Liquefied petroleum gas (LPG) or compressed natural gas (CNG) and used to power motor vehicles.

One of the most advanced reciprocating engines ever made was the 28-cylinder, 3,500 hp (2,600 kW) Pratt & Whitney R-4360 "Wasp Major" radial engine which powered the last generation of large piston-engined planes before the jet engine and turboprop took over from 1944 onward. It had a total engine capacity of 71.5 litres (2.52 cu ft). The largest reciprocating engine in production at present, but not the largest ever built, is the Wärtsilä-Sulzer RTA96-C turbocharged two-stroke diesel engine of 2006 built by Japan's Diesel United, Ltd. It is used to power the largest modern container ships such as the Emma Mærsk. It is five stories high (13.5 m/44 ft), 27 metres (89 ft) long, and weighs over 2,300 metric tons (2,500 short tons) in its largest 14 cylinders version producing more than 84.42 MW (114,800 bhp). Each cylinder has a capacity of 1,820 litres (64 cu ft), making a total capacity of 25,480 litres (900 cu ft) for the largest versions.

Engine Capacity

For piston engines, an engine's capacity is the engine displacement, in other words the volume swept by all the pistons of an engine in a single movement. It is generally measured in litres (L) or cubic inches (c.i.d. *or* cu in *or* in^3) for larger engines, and cubic centimetres (abbreviated cc) for smaller engines. All else being equal, engines with greater capacities are more powerful and provide greater torque at lower speed (rpm) and consumption of fuel increases accordingly, although power and fuel consumption are affected by many factors outside of engine displacement.

Other Modern Non-internal Combustion Types

Reciprocating engines that are powered by compressed air, steam or other hot gases are still used in some applications such as to drive many modern torpedoes or as pollution-free motive power. Most steam-driven applications use steam turbines, which are more efficient than piston engines. The French-designed Flow AIR vehicles use compressed air stored in a cylinder to drive a reciprocating engine in a pollution-free urban vehicle. Torpedoes may use a working gas produced by high test peroxide or Otto fuel II, which pressurise without combustion. The 230 kg (510 lb) Mark 46 torpedo, for example, can travel 11 km (6.8 mi) underwater at 74 km/h (46 mph) fuelled by Otto fuel without oxidant.

Miscellaneous Engines

Free-piston Engine

A free-piston engine is a linear, 'crankless' internal combustion engine, in which the piston motion is not controlled by a crankshaft but determined by the interaction of forces from the combustion chamber gases, a rebound device (e.g., a piston in a closed cylinder) and a load device (e.g. a gas compressor or a linear alternator).

The basic configuration of free-piston engines is commonly known as single piston, dual piston or opposed pistons, referring to the number of combustion cylinders. The free-piston engine is in practice restricted to the two-stroke operating principle, since a power stroke is required every fore-and-aft cycle.

First Generation

In 1807, de Rivaz tested an early engine that operated a free piston on a chain. The modern free-piston engine was proposed by R.P. Pescara and the original application was a single piston air compressor. The engine concept was a topic of much interest in the period 1930-1960, and a number of commercially available units were developed. These first generation free-piston engines were without exception opposed piston engines, in which the two pistons were mechanically linked to ensure symmetric motion. The free-piston engines provided some advantages over conventional technology, including compactness and a vibration-free design.

Air Compressors

The first successful application of the free-piston engine concept was as air compressors. In these engines, air compressor cylinders were coupled to the moving pistons, often in a multi-stage configuration. Some of these engines utilised the air remaining in the compressor cylinders to return the piston, thereby eliminating the need for a rebound device.

Free-piston air compressors were in use among others by the German Navy, and had the advantages of high efficiency, compactness and low noise and vibration.

Gas Generators

After the success of the free-piston air compressor, a number of industrial research groups started the development of free-piston gas generators. In these engines there is no load device coupled to the engine itself, but the power is extracted from an exhaust turbine. (The only load for the engine is supercharging the inlet air.)

A number of free-piston gas generators were developed, and such units were in widespread use in large-scale applications such as stationary and marine powerplants. Attempts were made to use free-piston gas generators for vehicle propulsion (e.g. in gas turbine locomotives) but without success.

Modern Applications

Modern applications of the free-piston engine concept include hydraulic engines, aimed for off-highway vehicles, and free-piston engine generators, aimed for use with hybrid electric vehicles.

Hydraulic

These engines are commonly of the single piston type, with the hydraulic cylinder acting as both load and rebound device using a hydraulic control system. This gives the unit high operational flexibility, and excellent part load performance has been reported for such engines.

Generators

The use of a free-piston engine with a linear generator is being investigated by a number of research groups, driven by the increasing interest in the hybrid electric vehicle concept in the automotive industry.

The first free piston generator was patented in 1959, and since then, a number of variations have been proposed. Examples include the Stelzer engine and the Free Piston Power Pack manufactured by Pempek Systems based on a German patent.

These engines are mainly of the dual piston type, giving a compact unit with high power-to-weight ratio. A challenge with this design is to find an electric machine with sufficiently low weight, and control challenges in the form of high cycle-to-cycle variations have been reported for dual piston engines.

Features and Potential Advantages

The operational characteristics of free-piston engines differ from those of conventional, crankshaft engines. The main difference is due to the piston motion not being restricted by a crankshaft in the free-piston engine, leading to the potentially valuable feature of variable compression ratio. This does, however, also present a control challenge, since the position of the dead centres must be accurately controlled in order to ensure fuel ignition and efficient combustion, and to avoid excessive in-cylinder pressures or, worse, the piston hitting the cylinder head.

Advantages

Potential advantages of the free-piston concept include:

- Simple design with few moving parts, giving a compact engine with low maintenance costs and reduced frictional losses.
- The operational flexibility through the variable compression ratio allows operation optimisation for all operating conditions and multi-fuel operation. The free-piston engine is further well suited for homogeneous charge compression ignition (HCCI) operation.
- High piston speed around top dead centre (TDC) and a fast power stroke expansion enhances fuel-air mixing and reduces the time available for heat transfer losses and the formation of temperature-dependent emissions such as nitrogen oxides (NOx).

Challenges

The main challenge for the free-piston engine is engine control, which can only be said to be fully solved for single piston hydraulic free-piston engines. Issues such as the influence of cycle-to-cycle variations in the combustion process and engine performance during transient operation in dual piston engines are topics that need further investigation.

Swing-piston Engine

A swing-piston engine is a type of internal combustion engine in which the pistons move in a circular motion inside a ring-shaped "cylinder", moving closer and further from each other to provide compression and expansion. Generally two sets of pistons are used, geared to move in a fixed relationship as they rotate around the cylinder. In some versions the pistons oscillate around a fixed centre, as opposed to rotating around the entire engine. The design has also been referred to as a oscillating piston engine, vibratory engine when the pistons oscillate instead of rotate, or toroidal engine based on the shape of the "cylinder".

The basic concept is very similar to the Wankel engine, the "traditional" rotary, but predates it by some time. It also shares a strong lineage with the free-piston engine. Like those designs, the swing-piston cylinder has to be open on one side to connect to the crankshaft, and is thus subject to the same sorts of sealing problems as other rotary designs. In addition, the swing-piston design retains a reciprocating motion and all the disadvantages that entails, namely the need for vibration damping and balancing. If the design oscillates around a fixed point, it also requires some sort of crank system to convert the motion into a rotary one. For these reasons, it has been said that the swing-piston design combines all the disadvantages of the rotary with all the disadvantages of the reciprocating.

Steam Engines

Swing-piston engines were initially introduced during the 1820s as alternate steam engine designs, prior to the widespread introduction of the steam turbine. In these examples the "piston" is typically not cylindrical as in a modern internal combustion design, and is generally rectangular in cross-section as seen from the top, rotating in a flat disk "cylinder". From the side they are either flat plates or pie-wedge shaped. The term "swing-piston" is not entirely accurate in these cases, but the operating cycle is identical and is properly considered here.

The first known example was introduced by Elijah Galloway in 1829 for ship propulsion. It featured a single vane rotating through 270 degrees. It appears this version was never built, although a model still exists in the Science Museum. Galloway also designed a wide variety of pure rotary engines using vanes as well.

A more serious attempt was the "Cambrian System" of John Jones in 1941. This design used two or three flat plates that were geared to move closer or further apart as the cycled continued. When the plates were at their closest point, steam was admitted between them using a valve, pushing them apart as the cycle continued. When the plates reached their maximum distance, an internal passage was uncovered that allowed the partially-expanded steam to flow across the centre of the device into the area on the other side of the vanes, which were now at their minimum distance. In this fashion the design was effectively a compound engine.

Many variations followed, and a number of these would see limited use in the field. Notable among them was John Ericsson's design of 1843, which powered the USS *Princeton*, the United States' first screw-powered steamship. Charles Parsons also examined the concept and appears to have produced two swing-piston engine designs before moving on to the steam turbine. The Roots brothers also designed a swing-piston engine of a unique type, although they are better known for their supercharger design.

Internal Combustion

It is unclear whether or not any internal combustion swing-piston engine has ever reached production, but the closest attempt appears to be the German World War II-era design by Otto Lutz. His design had six pistons in total, three each attached to two disks. The disks were geared to each other to form six chambers between the pistons, such that at any one time one set of three chambers were "close together" while the other set of three was "wide apart", varying between those two extremes as the disks rotated. The timing was arranged such that the chambers reached their "close together" point over the spark plug, and their "wide apart" point over the intake and exhaust ports. This action is similar to the Wankel, the primary difference being that Wankel creates compression and expansion via the shaping of the engine and rotor, as opposed to the relative motion of the pistons.

Lutz's engine was being designed as an experimental gas generator for a new type of aircraft engine, one that replaced a traditional centrifugal or axial compressor with his swing-piston design. Ultimately the exhaust would be used to drive a turbine, that power being used to drive a propeller to produce a turboprop. For this role the exhaust gas was too hot to be used directly in a turbine, given the available materials, so the engine featured a second "exhaust port" that vented cold pressurised air, which was then mixed into the hot exhaust. For direct power use, as opposed to driving a turbine, this "third area" of the engine can simply be left open to the air to avoid losing power to compression that will not be used.

The initial test engines had some minor problems, notably with sealing, but these were worked through and the engines were under test during 1944. One particularly nice feature of swing-piston engines is that they can be bolted back to back along a common crank shaft to make a larger engine, and with each additional stage the running becomes smoother and the only part that needs to be made larger is the crankshaft. A similar arrangement with a radial engine is generally more difficult to arrange, especially cooling, and ones with inline engine arrangements soon become so long that keeping the crankshaft from vibrating becomes a serious problem.

Each "cylinder" from Lutz's design was 0.70 m in diametre and only about 30 cm in depth, providing 445 hp from 140 kg, an excellent power-to-weight ratio compared even to jet engines of the era. A five-block version was proposed for his turboprop concept, providing 3,450 hp from an engine about 2 m long. While the power-to-weight was good, the *density* of the engine was simply superb.

The overall turboprop looked much more like a jet engine than a piston one. The swing-piston gas generator was located in the middle of a long nacelle, with a five-stage axial compressor in front and a three-stage turbine behind. The compressor was used both to act as a supercharger for the piston engine, as well as provide cold air to cool the turbine. The actual power to the propeller, combining both the pistons and the turbines, was 4,930 hp at 10,000 m altitude, far greater than any German wartime project.

Why all this complexity to produce a new version of an engine, the turboprop, whose primary advantage was simplicity? The main problem with conventional jet engines is that the combustion takes place in an open chamber, which is considerably less efficient than the closed chamber of a piston engine, where it has constant volume (or close to it). The Otto cycle or Diesel cycle used in piston engines has a much lower specific fuel consumption than the Brayton cycle of a traditional gas turbine engines at low speed. Lutz's design was intended to power very long-range bombers and patrol aircraft, where fuel economy was more important than simplicity and outright performance.

Lutz later patented the design under "Rotary compressor and other engines", United States Patent 2,301,667.

Other Examples

Lutz's design is not the only way to produce such an engine, BMW experimented with a traditional engine with poppet valves on the combustion chambers, which had been used a number of times previously in experiments. Another approach entirely is to recover some of the heat of the exhaust in a heat exchanger and use that instead of fuel to heat the compressed air, a concept used by General

Motors in a series of automobile turbines. Generally, however, improvements in the basic piston engine in "low power" roles have kept any of these advanced designs out of the marketplace.

More recently, starting in the 1990s, a number of inventors have re-introduced the concept as if it were new. Examples include Angel Labs' "Massive Yet Tiny" engine, the Rotoblock, the Roundengine, the Trochilic Engine and designs by Tschudi and Hoose. In 2009 Russian billionaire industrialist Mikhail Prokhorov announced his plans to enter an automotive business with a series of a lightweight hybrid vehicles utilising this design as their prime mover. Another recent introduction aimed at the hybrid market is the "Hüttlin Kugelmotor", which combines the swing-piston concept with a modified swashplate to produce a spherical design that directly powers an internal electrical generator.

IRIS Engine

The IRIS Engine is a design for a new type of internal combustion engine. Its inventors say that engines constructed using this design can be smaller, lighter and significantly more efficient than traditional engines of comparable horsepower and displacement. The design replaces the piston and cylinder architecture of conventional engines with a purportedly novel mechanism called the Internally Radiating Impulse Structure, or IRIS. In January 2008, the IRIS Engine design won first prize for transportation technology in NASA's annual "Create the Future" design competition. In October 2008, the Radial Expansion Engine (RXE), a variant of the IRIS design, won a major award in the ConocoPhillips Energy Prize competition.

Geometry of the IRIS

In an IRIS combustion chamber, a number of inverted segments of a circle, or "chordons," interact to create a continuously sealed chamber of variable volume. Instead of elongating during combustion, as a traditional engine does, the IRIS engine's chamber expands in diametre. The inventors claim that this innovation will reduce waste heat and will increase the amount of surface area the engine has available to produce torque.

Design Variations

IRIS engines are designed to run on traditional fuels, but could also be adapted to use biodiesel, natural gas, or hydrogen. IRIS chamber technology could also be utilised to create pumps, compressors and medical devices.

Designers

The IRIS was conceived of by Timber Dick, a Denver, Colorado inventor and businessman. Three of his sons, Corban, Levi, and Tomicah Tillemann-Dick, also contributed to the design and are credited on patent applications relating to the IRIS engine. The design recently won the $20,000 Dow Sustainability Award. A closely related design by the same team of inventors, the Radial Expansion Engine, was awarded a $75,000 prize in the ConocoPhillips Energy Prize Competition.

Bourke Engine

The Bourke Engine was designed by Russell Bourke in the 1920s, as an improved two-stroke engine. Despite finishing his design and building several working engines, the onset of World War II, lack of test results, and the poor health of his wife compounded to prevent his engine from ever coming successfully to market. The main claimed virtues of the design are that it has only two moving parts, is light weight, powerful, has two power pulses per revolution, and does not need oil mixed into the fuel.

Overview

The Bourke engine is basically a two-stroke design, with one horizontally opposed piston assembly using two pistons that move in the same direction at the same time, so that their operations are 180 degrees out of phase. The pistons are connected to a Scotch Yoke mechanism in place of the more usual crankshaft mechanism, thus the piston acceleration is perfectly sinusoidal. This causes the pistons to spend more time at top dead centre than conventional engines, allowing hydrogen detonation of the fuel to occur. The incoming charge is compressed in a chamber under the pistons, as in a conventional crankcase-charged two-stroke engine. The connecting-rod seal prevents the fuel from contaminating the bottom-end lubricating oil.

Operation

The operating cycle is very similar to that of a current production spark ignition two-stroke with crankcase compression, with two modifications:

1. The fuel is injected directly into the air as it moves through the transfer port.
2. The engine is designed to run without using spark ignition once it is warmed up. This is known as auto-ignition or dieseling,

and the air/fuel mixture starts to burn due to the high temperature of the compressed gas, and/or the presence of hot metal in the combustion chamber.

Design Features

The following design features have been identified.

Mechanical Features

- Scotch yoke instead of connecting rods to translate linear motion to rotary motion
- Fewer moving parts (only 2 moving assemblies per opposed cylinder pair) and the opposed cylinders are combinable to make 2, 4, 6, 8, 10, 12 or any even number of cylinders
- Smoother operation due to elimination of crank and slider mechanism
- The piston is connected to the Scotch yoke through a slipper bearing (a type of hydrodynamic tilting-pad fluid bearing)
- Mechanical fuel injection.
- Ports rather than valves.
- Easy maintenance (top overhauling) with simple tools.
- The Scotch yoke does not create lateral forces on the piston, reducing friction, vibration and piston wear.
- O-rings are used to seal joints rather than gaskets.
- The use of the Scotch Yoke reduces vibration from the motions of the connecting rod—for example, the peak acceleration in a Scotch yoke is 25% less than the acceleration in a conventional crank and slider arrangement. The piston movement and therefore vibration is sinusoidal so the engine could theoretically be perfectly counterbalanced, unlike a conventional engine which has harmonics in the piston movement courtesy of the lateral movement of the crankpin.
- The Scotch Yoke makes the pistons dwell very slightly longer at top dead centre, so the fuel burns more completely in a smaller volume.

Gas Flow and Thermodynamic Features

- Low exhaust temperature (below that of boiling water) so metal exhaust components are not required, plastic ones can be used if strength is not required from exhaust system

- Extremely fast hydrogen detonation burn time of the lean mixture so the engine can be considered to be a hydrogen detonation (i.e., explosion not deflagration) engine.
- 15:1 to 24:1 compression ratio for high efficiency and it can be easily changed as required by different fuels and operation requirements.
- Fuel is vapourised when it is injected into the transfer ports, and the turbulence in the intake manifolds and the piston shape above the rings stratifies the fuel air mixture into the combustion chamber.
- Lean burn for increased efficiency and reduced emissions.

Lubrication

- This design uses oil seals to prevent the pollution from the combustion chamber (created by piston ring blow-by in four-strokes and just combustion in two-strokes) from polluting the crankcase oil, extending the life of the oil as it is used slowly for keeping the rings full of oil to hold and use to lubricate. Oil was shown to be used slowly by the dropfull as needed, but checking the quantity and cleanness of it was still recommended by Russell Bourke, its creator.
- The lubricating oil in the base is protected from combustion chamber pollution by an oil seal over the connecting rod.
- The piston rings are supplied with oil from a small supply hole in the cylinder wall at bottom dead centre.

Claimed and Measured Performance

- Efficiency 0.25 (lb/h)/hp is claimed - about the same as the best diesel engine, or roughly twice as efficient as the best two strokes. This is equivalent to a thermodynamic efficiency of 55.4%, which is an exceedingly high figure for a small internal combustion engine. In a test witnessed by a third party, the actual fuel consumption was 1.1 hp/(lb/h), or 0.9 (lb/h)/hp, equivalent to a thermodynamic efficiency of about 12.5%, which is typical of a 1920s steam engine.
- Power to weight 0.9 to 2.5 hp/lb is claimed, although no independently witnessed test to support this has been documented. The upper range of this is roughly twice as good as the best four-stroke production engine shown here, or 0.1 hp/

lb better than a Graupner G58 two-stroke. The lower claim is unremarkable, easily exceeded by production four-stroke engines, never mind two strokes.

- Emissions Achieved virtually no hydrocarbons (80 ppm) or carbon monoxide (less than 10 ppm) in published test results, however no power output was given for these results, and NOx was not measured.
- Low Emissions The engine is claimed to be able to operate on hydrogen or any hydro-carbon fuel without any modifications, producing only water vapour and carbon dioxide as emissions.

Engineering Critique of the Bourke Engine

The Bourke Engine has some interesting features, but the extravagant claims for its performance are unlikely to be borne out by real tests. Many of the claims are contradictory.

1. Seal friction from the seal between the air compressor chamber and the crankcase, against the connecting rod, will reduce the efficiency.
2. Efficiency will be reduced due to pumping losses, as the air charge is compressed and expanded twice but energy is only extracted for power in one of the expansions per piston stroke.
3. Engine weight is likely to be high because it will have to be very strongly built to cope with the high peak pressures seen as a result of the rapid high temperature combustion.
4. Each piston pair is highly imbalanced as the two pistons move in the same direction at the same time, unlike in a boxer engine. This will limit the speed range and hence the power of the engine, and increase its weight due to the strong construction necessary to react the high forces in the components.
5. High speed two-stroke engines tend to be inefficient compared with four-strokes because some of the intake charge escapes unburnt with the exhaust.
6. When the charge is transferred from the compressor chamber to the combustion chamber it will cool down, reducing the efficiency of the engine.
7. Use of excess air will reduce the torque available for a given engine size.

8. Forcing the exhaust out rapidly through small ports will incur a further efficiency loss.
9. Operating an internal combustion engine in detonation reduces efficiency due to heat lost from the combustion gases being scrubbed against the combustion chamber walls by the shock waves.
10. Emissions - although some tests have shown low emissions in some circumstances, these were not necessarily at full power. As the scavenge ratio (i.e. engine torque) is increased more HC and CO will be emitted.
11. Increased dwell time at TDC will allow more heat to be transferred to the cylinder walls, reducing the efficiency.
12. When running in auto-ignition mode the timing of the start of the burn is controlled by the operating state of the engine, rather than directly as in a spark ignition or diesel engine. As such it may be possible to optimise it for one operating condition, but not for the wide range of torques and speeds that an engine typically sees. The result will be reduced efficiency and higher emissions.
13. If the efficiency is high, then combustion temperatures must be high, as required by the Carnot cycle, and the air fuel mixture must be lean. High combustion temperatures and lean mixtures cause nitrogen dioxide to be formed.

Chapter 2

Two-stroke Engine

Figure 1: Brons two-stroke V8 Diesel engine driving a Heemaf generator.

A two-stroke engine is an internal combustion engine that completes the process cycle in one revolution of the crankshaft (an up stroke and a down stroke of the piston, compared to twice that number for a four-stroke engine). This is accomplished by using the end of the combustion stroke and the beginning of the compression stroke to perform simultaneously the intake and exhaust (or scavenging) functions. In this way, two-stroke engines often provide high specific power, at least in a narrow range of rotational speeds. The functions of some or all of the valves required by a four-stroke engine are usually served in a two-stroke engine by ports that are opened and closed by the motion of the piston(s), greatly reducing the number of moving parts. Gasoline (spark ignition) versions are particularly useful in lightweight (portable) applications, such as chainsaws, and the concept is also used in diesel compression ignition engines in large and weight insensitive applications, such as ships and locomotives.

The first commercial two-stroke engine involving in-cylinder compression is attributed to Scottish engineer Dugald Clerk, who in 1881 patented his design, his engine having a separate charging cylinder. The crankcase-scavenged engine, employing the area below the piston as a charging pump, is generally credited to Englishman Joseph Day.

Applications

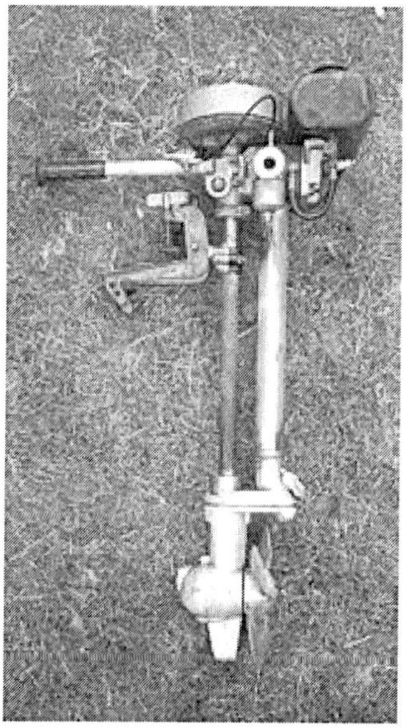

Figure 2: Lateral view of a two-stroke Forty series British Seagull outboard engine, the serial number dates it to 1954/1955.

The two-stroke engine was very popular throughout the 20th century in motorcycles and small-engined devices, such as chainsaws and outboard motors, and was also used in some cars, a few tractors and many ships. Part of their appeal was their simple design (and resulting low cost) and often high power-to-weight ratio. Because of the lower cost to rebuild and maintain, the two stroke engine was incredibly popular in the dirt bike and motocross industry for many years, until recently when the EPA required the industry to switch to four-stroke engines because they emit less pollution than two stroke engines.

Many designs use total-loss lubrication, with the oil being burned in the combustion chamber, causing "blue smoke" and other types of exhaust pollution. This is a major reason for two-stroke engines being replaced by four-stroke engines in many applications.

Two-stroke engines continue to be commonly used in high-power, handheld applications such as string trimmers and chainsaws. The light overall weight, and light-weight spinning parts give important operational and even safety advantages. For example, only a two-stroke engine that uses a gasoline-oil mixture can power a chainsaw operating in any position.

These engines are still used for small, portable, or specialised machine applications such as outboard motors, high-performance, small-capacity motorcycles, mopeds, underbones, scooters, tuk-tuks, snowmobiles, karts, ultralights, model aeroplanes (and other model vehicles) and lawnmowers. The two-stroke cycle is used in many diesel engines, most notably large industrial and marine engines, as well as some trucks and heavy machinery.

A number of mainstream automobile manufacturers have used two-stroke engines in the past, including the Swedish Saab and German manufacturers DKW and Auto-Union. The Japanese manufacturer Suzuki did the same in the 1970s. Production of two-stroke cars ended in the 1980s in the West, but Eastern Bloc countries continued until around 1991, with the Trabant and Wartburg in East Germany. Lotus of Norfolk, UK, has a prototype direct-injection two-stroke engine intended for alcohol fuels called the Omnivore which it is demonstrating in a version of the Exige.

Different Two-stroke Design Types

Although the principles remain the same, the mechanical details of various two-stroke engines differ depending on the type. The design types vary according to the method of introducing the charge to the cylinder, the method of scavenging the cylinder (exchanging burnt exhaust for fresh mixture) and the method of exhausting the cylinder.

Piston-controlled Inlet Port

Piston port is the simplest of the designs. All functions are controlled solely by the piston covering and uncovering the ports as it moves up and down in the cylinder. A fundamental difference from typical four-stroke engines is that the crankcase is sealed and forms part of the induction process in gasoline and hot bulb engines. Diesel engines have mostly a Roots blower or piston pump for scavenging.

Reed Inlet Valve

The reed valve is a simple but highly effective form of check valve commonly fitted in the intake tract of the piston-controlled port. They allow asymmetric intake of the fuel charge, improving power and economy, while widening the power band. They are widely used in ATVs and marine outboard engines.

Rotary Inlet Valve

The intake pathway is opened and closed by a rotating member. A familiar type sometimes seen on small motorcycles is a slotted disk attached to the crankshaft which covers and uncovers an opening in the end of the crankcase, allowing charge to enter during one portion of the cycle. Another form of rotary inlet valve used on two-stroke engines employs two cylindrical members with suitable cutouts arranged to rotate one within the other - the inlet pipe having passage to the crankcase only when the two cutouts coincide. The crankshaft itself may form one of the members, as in most glow plug model engines. In another embodiment, the crank disc is arranged to be a close-clearance fit in the crankcase, and is provided with a cutout which lines up with an inlet passage in the crankcase wall at the appropriate time, as in the Vespa motor scooter.

The advantage of a rotary valve is it enables the two-stroke engine's intake timing to be asymmetrical, which is not possible with piston port type engines. The piston port type engine's intake timing opens and closes before and after top dead centre at the same crank angle, making it symmetrical, whereas the rotary valve allows the opening to begin earlier and close earlier.

Rotary valve engines can be tailored to deliver power over a wider speed range or higher power over a narrower speed range than either piston port or reed valve engine. Where a portion of the rotary valve is a portion of the crankcase itself, it is particularly important that no wear is allowed to take place.

Crossflow-scavenged

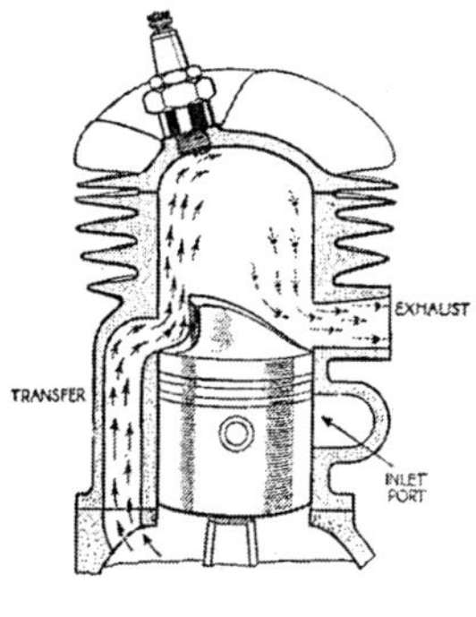

Figure 3: Deflector piston with crossflow scavenging

In a crossflow engine, the transfer and exhaust ports are on opposite sides of the cylinder, and a deflector on the top of the piston directs the fresh intake charge into the upper part of the cylinder, pushing the residual exhaust gas down the other side of the deflector and out the exhaust port. The deflector increases the piston's weight and exposed surface area, and also makes it difficult to achieve an efficient combustion chamber shape. This design has been largely superseded by the loop scavenging method (below), although for smaller or slower engines, the crossflow-scavenged design can be an acceptable approach.

Schnuerle Porting

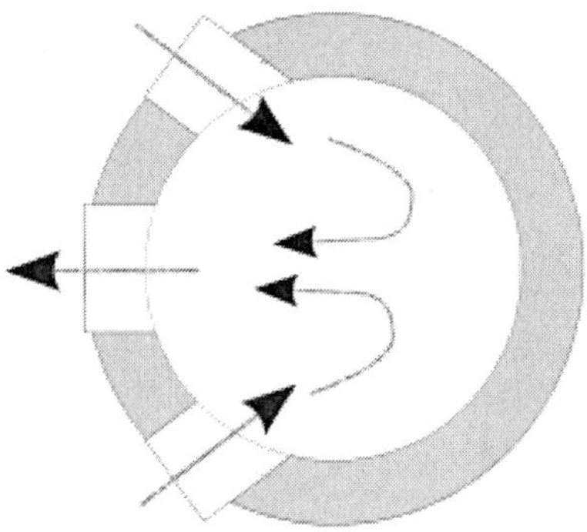

Figure 4: Cylinder and ports, viewed from above

Schnuerle porting is a system to improve efficiency of a valveless two-stroke engine. The intake and exhaust ports cut in the cylinder wall are shaped to give a more efficient transfer of intake and exhaust gases.

Description

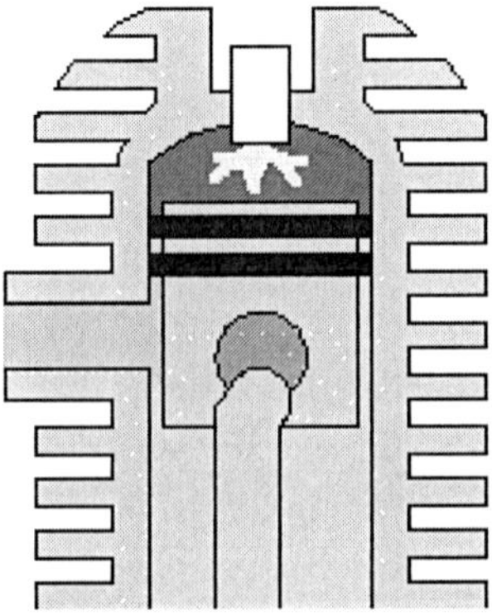

Figure 5: Cylinder scavenging, from the side

Gas flow within the two-stroke engine is even more critical than for a four-stroke engine, as the two flows are both entering and leaving the combustion chamber simultaneously. A well-defined flow pattern is required, avoiding any turbulent mixing. The efficiency of the two-stroke engine depends on effective scavenging, the complete replacement of the old spent charge with a fresh charge.

Apart from large diesels with separate superchargers, two-stroke engines are generally piston-ported and use their crankcase beneath the piston for compression. The cylinder has a transfer port (inlet from crankcase to combustion chamber) and an exhaust port cut into it. These are opened together, as the piston moves downwards past them.

The simplest arrangement is a single transfer and single exhaust port, opposite each other. This "cross scavenging" performs poorly, as there is tendency for the flow to pass from the inlet directly to the exhaust, wasting some of the fuel mixture and also poorly scavenging the upper part of the chamber. Before Schnuerle porting, a deflector on top of the piston was used to direct the gas flow from the transfer port upwards, in a U-shapped loop around the combustion chamber roof and then down and out through the exhaust port. Apart from the gas flow never quite following this ideal path and tending to mix instead, this also gave a poorly shaped combustion chamber with long, thin flame paths.

In 1926, the German engineer Adolf Schnürle developed the system of ports that bears his name. The ports were relocated to both be on the same side of the cylinder, with the transfer port being split into two angled ports, one on either side of the exhaust port. A deflector piston was no longer required. The gas flow was now a circular loop, flowing in and across the piston crown from the transfer ports, up and around the combustion chamber and then out through the exhaust port. With Schnuerle porting, the piston crown may be of any shape, even bowl shaped. This permits a far better combustion chamber shape and flame path, giving better combustion, particularly at high speeds.

Loop Scavenging

As Schnuerle porting encourages flow in a loop, it is termed "loop scavenging". Historically, the deflector piston form of cross scavenging was termed "loop scavenging", after the supposed shape of the flow. Schnuerle flow was termed "reverse loop scavenging". As the first of these was realised to be inaccurate, the later form adopted the simpler name. These original terms are now obsolete and no longer used.

Adolf Schnürle

The system is named after its inventor, Adolf Schnürle. Either "Schnürle" or the more common Anglicisation as "Schnuerle" are generally acceptable. It also appears as "Schnürrle" , but "Schneurle" is a mis-spelling.

Adolf Schnuerle was a prolific engineer and is named on many Patent Documents.

Uniflow-scavenged

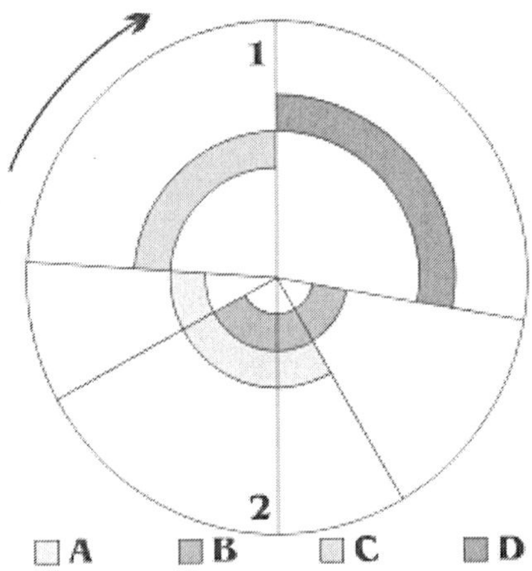

Figure 6: The Uniflow Two-stroke cycle

1=TDCdg

2=Bbb

A: Intake (effective scavenging H"140°-250°)

B: Exhaust

C: Compression

D: Expansion (power)

In a uniflow engine, the mixture, or air in the case of a diesel, enters at one end of the cylinder controlled by the piston and the exhaust exits at the other end controlled by an exhaust valve or piston. The scavenging gas-flow is therefore in one direction only, hence the name uniflow. The valved arrangement is common in diesel locomotives (Electro-Motive Diesel) and large marine two-stroke engines (Wärtsilä). Ported types are represented by the opposed piston design in which there are two pistons in each cylinder, working in opposite directions such as the Junkers Jumo and Napier Deltic. The once-popular split-single design falls into this class, being effectively a folded uniflow. With advanced angle exhaust timing, uniflow engines can be supercharged with a crankshaft-driven (piston or Roots) blower.

The latest invention, called the Reversed Uniflow two-stroke engine, has a large intake valve for compressed intake air without fuel-oil mixture. Direct fuel injection is to be used for gasoline or diesel fuel, pending intake air pressure. This engine will work on the Miller cycle. US Patent #6889636.

Stepped Piston Engine

The piston of this engine is "top-hat" shaped; the upper section forms the regular cylinder, and the lower section performs a scavenging function. The units run in pairs, with the lower half of one piston charging an adjacent combustion chamber. This system is still partially dependent on total loss lubrication (for the upper part of the piston), the other parts being sump lubricated with cleanliness and reliability benefits. The piston weight is only about 20% heavier than a loop-scavenged piston because skirt thicknesses can be less. The patents on this design are held by Bernard Hooper Engineering Ltd (BHE).

Power Valve Systems

Many modern two-stroke engines employ a power valve system. The valves are normally in or around the exhaust ports. They work in one of two ways: either they alter the exhaust port by closing off the top part of the port, which alters port timing, such as Ski-doo R.A.V.E, Yamaha YPVS, Honda RC-Valve, Cagiva C.T.S. or Suzuki AETC systems, or by altering the volume of the exhaust, which changes the resonant frequency of the expansion chamber, such as the Honda V-TACS system. The result is an engine with better low-speed power without sacrificing high-speed power.

Gasoline Direct Injection

In internal combustion engines, gasoline direct injection (GDI), also known as petrol direct injection or direct petrol injection, is a variant of fuel injection employed in modern two-stroke and four-stroke gasoline engines. The gasoline is highly pressurised, and injected via a common rail fuel line directly into the combustion chamber of each cylinder, as opposed to conventional multi-point fuel injection that happens in the intake tract, or cylinder port.

In some applications, gasoline direct injection enables a stratified fuel charge (ultra lean burn) combustion for improved fuel efficiency, and reduced emission levels at low load.

Figure 7: Combustion chamber of a 3.5 Ford EcoBoost engine

Theory of Operation

The major advantages of a GDI engine are increased fuel efficiency and high power output. Emissions levels can also be more accurately controlled with the GDI system. The cited gains are achieved by the precise control over the amount of fuel and injection timings that are varied according to the load conditions. In addition, there are no throttling losses in some GDI engines, when compared to a conventional fuel injected or carbureted engine, which greatly improves efficiency, and reduces 'pumping losses' in engines without a throttle plate. Engine speed is controlled by the engine control unit/engine management system (EMS), which regulates fuel injection function and ignition timing, instead of having a throttle plate that restricts the incoming air supply. Adding this function to the EMS requires considerable enhancement of its processing and memory, as direct injection plus the engine speed management must have very precise algorithms for good performance and drivability.

The engine management system continually chooses among three combustion modes: ultra lean burn, stoichiometric, and full power output. Each mode is characterised by the air-fuel ratio. The stoichiometric air-fuel ratio for gasoline is 14.7:1 by weight, but ultra lean mode can involve ratios as high as 65:1 (or even higher in some engines, for very limited periods). These mixtures are much leaner than in a conventional engine and reduce fuel consumption considerably.

- Ultra lean burn mode is used for light-load running conditions, at constant or reducing road speeds, where no acceleration is required. The fuel is not injected at the intake stroke but rather at the latter stages of the compression stroke, so that the small amount of air-fuel mixture is optimally placed near the spark plug. This stratified charge is surrounded mostly by air, which keeps the fuel and the flame away from the cylinder walls for lowest emissions and heat losses. The combustion takes place in a toroidal (donut-shaped) cavity on the piston's surface. The cavity is displaced to one side of the piston, the side that has the fuel injector. This technique enables the use of ultra-lean mixtures that would be impossible with carburetors or conventional fuel injection.

- Stoichiometric mode is used for moderate load conditions. Fuel is injected during the intake stroke, creating a homogenous fuel-air mixture in the cylinder. From the stoichiometric ratio, an optimum burn results in a clean exhaust emission, further cleaned by the catalytic converter.
- Full power mode is used for rapid acceleration and heavy loads (as when climbing a hill). The air-fuel mixture is homogenous and the ratio is slightly richer than stoichiometric, which helps prevent knock (pinging). The fuel is injected during the intake stroke.

Direct injection may also be accompanied by other engine technologies such as variable valve timing (VVT) and tuned/multi path or variable length intake manifolding (VLIM, or VIM). Water injection or (more commonly) exhaust gas recirculation (EGR) may help reduce the high nitrogen oxides (NOx) emissions that can result from burning ultra lean mixtures.

It is also possible to inject more than once during a single cycle. After the first fuel charge has been ignited, it is possible to add fuel as the piston descends. The benefits are more power and economy, but certain octane fuels have been seen to cause exhaust valve erosion. For this reason, most companies have ceased to use the Fuel Stratified Injection (FSI) operation during normal running.

Tuning up an early generation FSI power plant to generate higher power is difficult, since the only time it is possible to inject fuel is during the induction phase. Conventional injection engines can inject throughout the 4-stroke sequence, as the injector squirts onto the back of a closed valve. A direct injection engine, where the injector injects directly into the cylinder, is limited to the suction stroke of the piston. As the RPM increases, the time available to inject fuel decreases. Newer FSI systems that have sufficient fuel pressure to inject even late in compression phase do not suffer to the same extent;

however, they still do not inject during the exhaust cycle (they could but it would just waste fuel). Hence, all other factors being equal, an FSI engine needs higher-capacity injectors to achieve the same power as a conventional engine.

History—Early systems

The first use of direct gasoline injection was on the Hesselman engine invented by Swedish engineer Jonas Hesselman in 1925. Hesselman engines used the ultra lean burn principle and injected the fuel in the end of the compression stroke and then ignited it with a spark plug, it was often started on gasoline and then switched over to run on diesel or kerosene.

The Hesselman engine was a low compression design constructed to run on heavy fuel oils. Direct gasoline injection was used on production aircraft during WWII, with German (Junkers Jumo 210, Daimler-Benz DB 601, both 1937), Soviet (Shvetsov ASh-82FN, 1943, Chemical Automatics Design Bureau - KB Khimavtomatika) and US (Wright R-3350, 1944) designs. The first automotive direct injection system used to run on gasoline was developed by Bosch, and was introduced by Goliath and Gutbrod in 1952. The 1955 Mercedes-Benz 300SL, the first production sports car to use fuel injection, used direct injection. The Bosch fuel injectors were placed into the bores on the cylinder wall used by the spark plugs in other Mercedes-Benz six-cylinder engines (the spark plugs were relocated to the cylinder head). Later, more mainstream applications of fuel injection favoured the less-expensive indirect injection methods.

In the early-1970s, research was conducted with the backing of American Motors Corporation (AMC) to develop a Straticharge Continuous Fuel-Injection (SCFI) system. The conventional spark ignited internal combustion AMC straight-6 engine was modified with a redesigned cylinder head. The system incorporated a mechanical device that automatically responded to the engine's airflow and loading conditions with two separate fuel-control pressures supplied to two sets of continuous-flow injectors. Flexibility was designed into the SCFI system for trimming it to a particular engine. Prototype "straticharge" engine road testing was performed using a 1973 AMC Hornet, but the mechanical fuel controls had teething problems.

During the late-1970s, the Ford Motor Company developed a stratified-charge engine they called "ProCo" (programmed combustion), utilising a unique high-pressure pump and direct injectors. At least one hundred and fifteen (115) Crown Victoria cars were built at Ford's

Atlanta Assembly in Hapeville, Georgia using a ProCo V8 engine. The project was cancelled for several reasons: electronic controls, a key element, were in their infancy; pump and injector costs were extremely high; and lean combustion produced nitrogen oxides in excess of near future United States Environmental Protection Agency (EPA) limits. The three-way catalytic converter proved to be a less expensive solution.

Later Systems

In 1996 gasoline direct injection reappeared in the automotive market. Mitsubishi was the first with a GDI engine in the Japanese market with its Galant/Legnum's *4G93* 1.8 L inline-four. It was subsequently brought to Europe in 1997 in the Carisma, although Europe's then high-sulphur unleaded fuel led to emissions problems, and fuel efficiency was less than expected. It also developed the first six-cylinder GDI powerplant, the *6G74* 3.5 L V6, in 1997. Mitsubishi applied this technology widely, producing over one million GDI engines in four families by 2001.

In 1997 Nissan released the Leopard featuring the VQ30DD equipped with direct injection.

In 1998, Toyota's D4 direct injection system first appeared on various Japanese market vehicles equipped with the *SZ* and *NZ* engines. Toyota later introduced its D4 system to European markets with the *1AZ-FSE* engine found in the 2001 Avensis. and US markets in 2005 with the *3GR-FSE* engine found in the Lexus GS 300. Toyota's *2GR-FSE* V6 uses a more advanced direct injection system, which combines both direct and indirect injection using two fuel injectors per cylinder, a traditional port fuel injector (low pressure) and a direct fuel injector (high-pressure) in a system known as D4-S.

In 1999, Renault introduced the 2.0 IDE (Injection Direct Essence), first on the Megane. Rather than following the lean burn approach, Renault's design uses high ratios of exhaust gas recirculation to improve economy at low engine loads, with direct injection allowing the fuel to be concentrated around the spark. Later gasoline direct injection engines have been tuned and marketed for their high performance as well as increased fuel efficiency. PSA Peugeot Citroën, Hyundai and Volvo licensed Mitsubishi's GDI technology in 1999. Although other companies have since developed gasoline direct injection engines, the acronym 'GDI' (with an uppercase final "I") remains a registered trademark of Mitsubishi Motors.

In 2000, the Volkswagen Group introduced its gasoline direct injection engine in the Volkswagen Lupo, a 1.4 L inline-four unit, under the product name "Fuel Stratified Injection" (FSI). The technology was adapted from Audi's Le Mans prototype race car R8. Volkswagen Group marques use direct injection in its 2.0 L FSI turbocharged and naturally aspirated four-cylinder engines. Later, a 2.0 L inline-four unit was introduced in the model year 2003 Audi A4. PSA Peugeot Citroën introduced its first GDi (HPi) engine in 2000 in the Citroën C5 and Peugeot 406. It was a 2.0-litre 16-valve EW10 D unit with 140 hp (104 kW), the system was licensed from Mitsubishi.

In 2002, the Alfa Romeo 156 with a direct-injection engine, the JTS (Jet Thrust Stoichiometric) went on sale and today the technology is used on almost every Alfa Romeo engine.

In 2003, A new 1.8 L Duratec SCi naturally aspirated engine made its production debut in the Ford Mondeo in 2003. Ford introduced its first European Ford engine to use direct injection technology in 2001, badged SCi (Smart Charge injection) for Direct-Injection-Spark-Ignition (DISI). The range will include some turbocharged derivatives, including the 1.1 L, three-cylinder turbocharged unit showcased at the 2002 Geneva Show.

In 2003, BMW introduced a low-pressure gasoline direct injection N73 V12. This initial BMW setup could not enter lean-burn mode, but the company introduced its second-generation High Precision Injection (HPI) system on the updated N52 straight-6 in 2006, which used high-pressure injectors. This system surpasses many others with a wider envelope of lean-burn time, increasing overall efficiency. PSA is cooperating with BMW on a new line of engines that made its first appearance in the 2007 MINI Cooper S. Honda released their own direct injection system on the Stream sold in Japan. Honda's fuel injector is placed directly atop the cylinder at a 90-degree angle rather than a slanted angle.

Since 2004, General Motors has released three such direct injected engines: in 2004, a 155 hp (116 kW) version of the 2.2 L Ecotec used in the Opel/Vauxhall Vectra and Signum in 2005, a 2.0 L turbocharged Ecotec for the new Opel GT, Pontiac Solstice GXP, and the Saturn Sky Red Line, in 2007 the same engine was used in the Super Sport versions of the Chevrolet Cobalt and the HHR. Also in 2007, the 3.6 L LLT became available in the redesigned Cadillac CTS and STS. The 3.6 L was added to the 2009 model GMC Acadia, Chevrolet Traverse, Saturn Outlook, Buick Enclave, and the 2010 Chevy Camaro. In 2004 Isuzu produced the first GDi engine sold in a mainstream American vehicle, standard on the 2004 Axiom and optional on the 2004 Rodeo.

Isuzu claimed the benefit of GDi is that the vapourising fuel has a cooling effect, allowing a higher compression ratio (10.3:1 versus 9.1:1) that boosts output by 20 hp (15 kW), and that 0-to-60 mph times drop from 8.9 to just 7.5 seconds, with the quarter-mile being cut from 16.5 to 15.8 seconds.

In 2005, Mazda began to use their own version of direct-injection in the Mazdaspeed6 and later on the CX-7 sport-utility, and the new Mazdaspeed3 in the US and European market. It is referred to as Direct Injection Spark Ignition (DISI).

In 2006, BMW released the new N54 twin-turbo-charged direct injection inline-six engine for its 335i Coupe and later for the 335i Sedan, 535i series and the 135i models. Mercedes-Benz released its direct injection system (Charged Gasoline Injection, or "CGI") on the CLS 350 CGI featuring common rail, piezo-electric direct fuel injectors. The CLS 350 CGI offers 292 BHP versus 272 BHP for the CLS 350, with reduced carbon dioxide emissions and improved fuel economy.

In 2007, Ford introduced its new Ford EcoBoost engine technology designed for a range of global vehicles (from small cars to large trucks). The engine first appeared in the 2007 Lincoln MKR Concept under the name *TwinForce*. The new global EcoBoost family of 4-cylinder and 6-cylinder engines features turbocharging and direct injection technology (GTDI - Gasoline Turbocharged Direct Injection). A 2.0 L version was unveiled in the 2008 Ford Explorer America Concept.

In 2008, BMW released the X6 xDrive50i equipped with a direct injected twin turbo N63 V8 engine.

In 2009, Ferrari began selling the front-engine California with a direct injection system, and announced that its new 458 Italia car will also feature a direct injection system, a first for Ferrari mid-rear engine setups. Porsche also began selling the 997 and Cayman equipped with direct injection. Ford produced the new generation Taurus SHO and Flex with a 3.5 L twin-turbo EcoBoost V-6 with direct injection. Holden has also added two direct injection engines as standard on the V6 variant Commodores under the name of SIDI or Spark Ignition Direct Injection. The Infiniti Essence concept car is powered by a direct injected twin turbo V6. The Jaguar Land Rover AJ-V8 Gen III 5.0 L engine (introduced in August 2009 for the 2010 model year) features spray-guided direct injection.

In 2010 Infiniti will produce the M56 which includes DI. Motus Motorcycles is developing, with Katech Engines, a direct-injected V4 engine named the KMV4 as the powertrain for their MST motorcycles. The Hyundai Sonata 2011 model will come with GDI engines. Hyundai's Theta I-4 engine family is a proprietary design, engineered in Namyang, Korea and currently in production for applications all over the world.

In Two-stroke Engines

The benefits of direct injection are even more pronounced in two-stroke engines, because it eliminates much of the pollution they cause. In all two-strokes, the exhaust and intake ports are both open at the same time, at the bottom of the piston stroke, for "scavenging". In coventional two-strokes, a large portion of the fuel/air mixture entering the cylinder from the crankcase through the intake ports goes directly out, unburned, through the exhaust port. With direct injection, only air (and usually some oil) comes from the crankcase, and fuel is not injected until the piston rises and all ports are closed.

Some Goliath two stroke cars built in the early 1950s had direct injection, but their engines were soon superseded by four strokes.

Two types of GDi are used in two-strokes: low-pressure air-assisted, and high-pressure. The former, developed by Orbital Engine Corporation of Australia (now Orbital Corporation) injects a mixture of fuel and compressed air into the combustion chamber. When the air expands it atomizes the fuel. The Orbital system is used in motor scooters manufactured by Aprilia, Piaggio, Peugeot and Kymco, in outboard motors manufactured by Mercury and Tohatsu, and in personal watercraft manufactured by Bombardier Recreational Products (BRP).

The high-pressure direct injector for two stroke engines was developed in the early 1990s by Ficht GmbH of Kirchseeon Germany. Outboard Marine Corporation (OMC) licensed the technology in 1995 and introduced it on a production outboard engine in 1996. OMC purchased a controlling interest in Ficht in 1998. Beset by extensive warranty claims for its Ficht outboards and prior and concurrent management-financial problems, OMC declared bankruptcy in December 2000 and the engine manufacturing portion and brands (Evinrude Outboard Motors and Johnson Outboards), including the Ficht technology, were purchased by BRP in 2001.

Evinrude introduced the E-Tec system, an improvement to the Ficht fuel injection, in 2003, based on U.S. patent 6,398,511. In 2004, Evinrude received the EPA Clean Air Excellence Award for their outboards utilising the E-Tec system. The E-Tec system has recently also been adapted for use in performance two-stroke snowmobiles.

Yamaha also has a high-pressure direct injection (HPDI) system for two-stroke outboards. It differs from the Ficht/E-Tec and Orbital direct injection systems because it uses a separate, belt driven, high-pressure, mechanical fuel pump to generate the pressure necessary for injection in a closed chamber. This is similar to most current 4-stroke automotive designs. EnviroFit, a non-profit corporation sponsored by Colorado State University, has developed direct injection retrofit kits for two-stroke motorcycles in a project to reduce air pollution in Southeast Asia, using technology developed by Orbital Corporation of Australia. The World Health Organisation says air pollution in Southeast Asia and the Pacific causes 537,000 premature deaths each year. The 100-million two-stroke taxis and motorcycles in that part of the world are a major cause.

Future—Twin-fuel Engines

Code named Bobcat, the new twin-fuel engine from Ford is based on a 5.0L V8 engine block but uses E85 cylinder injection and gasoline port injection. The engine was co-developed with Ethanol Boosting Systems, LLC of Cambridge, Massachusetts, which calls its trademarked process DI Octane Boost. The direct injection of ethanol increases the octane of regular gasoline from 88-91 octane to more than 150 octane. The Bobcat project was unveiled to the United States Department of Energy and the SAE International in April 2009.

Formula One

As part of the rule changes under discussion for the 2013 season, GDI has been mentioned as a potential technology of interest by Ferrari.

Two-stroke Diesel Engine

Figure 8: Brons two stroke V8 Diesel engine driving a Heemaf generator.

A two stroke diesel is a diesel engine that works in two strokes. A diesel engine is an internal combustion engine which operates using the Diesel cycle. Invented in 1892 by German engineer Rudolf Diesel, it was based on the hot bulb engine design and patented on February 23, 1893. During the period of 1900 to 1930, four-stroke diesel engines enjoyed a relative dominance in practical diesel applications. Charles F. Kettering and colleagues, working at the various incarnations of Electro-Motive and at the General Motors Research Corporation during the 1930s, advanced the art and science of two-stroke diesel technology to yield engines with much higher power-to-weight ratios than the two-stroke diesels of old. This work was instrumental in bringing about the dieselisation of railroads in the 1940s and 1950s.

All diesel engines use compression ignition, a process by which fuel is injected after the air is compressed in the combustion chamber causing the fuel to self ignite. By contrast, gasoline engines utilise the Otto cycle, in which fuel and air are mixed before entering the combustion chamber and then ignited by a spark plug.

Two Strokes

Two-stroke internal combustion engines are more simple mechanically than four-stroke engines, but more complex in thermodynamic and aerodynamic processes. In a two-stroke engine, the four "cycles" of internal combustion engine theory (intake, compression, ignition, exhaust) occur in one revolution, while in a four-stroke engine it occurs in two complete revolutions. In a two-stroke engine, more than one function occurs at any given time during the engine's operation.

- *Intake* begins when the piston is near the bottom dead centre. Air is admitted to the cylinder through ports in the cylinder wall (there are no intake valves). All two-stroke Diesel engines require artificial aspiration to operate, and will either use a mechanically-driven blower or a hybrid turbo-supercharger to charge the cylinder with air. In the early phase of intake, the air charge is also used to force out any remaining combustion gases from the preceding power stroke, a process referred to as scavenging.
- As the piston rises, the intake charge of air is compressed. Near top dead centre, fuel is injected, resulting in combustion due to the extremely high pressure and heat created by compression, which drives the piston downward. As the piston

moves downward in the cylinder it will reach a point where the exhaust port is opened to expel the high-pressure combustion gasses. However, most current two-stroke diesel engines use top-mounted poppet valves and uniflow scavenging. Continued downward movement of the piston will expose the air intake ports in the cylinder wall, and the cycle will start again.

Roots Blower

The Roots blower is commonly used on the two stroke diesel engine, which requires some form of forced induction. In this application, the blower does not provide significant compression and these engines are considered naturally aspirated; turbochargers are generally used when significant "boost" is needed.

Lubrication

Most small petrol two-stroke engines cannot be lubricated by oil contained in their crankcase and sump, since the crankcase is already being used to pump fuel-air mixture into the cylinder. Traditionally, the moving parts (both rotating crankshaft and sliding piston) were lubricated by a premixed fuel-oil mixture (at a ratio between 16:1 and 50:1). As late as the 1960s, petrol stations would often have a separate pump to deliver such a premix fuel to motorcycles. Even then, in many cases, the rider would carry a bottle of their own two-stroke oil. Taking care to close the fuel-tap first, he or she would metre in a little oil (using the cap of the bottle) and then put in the petrol, this action mixing the two liquids.

Modern two-stroke engines pump lubrication from a separate tank of oil. This is still a total-loss system with the oil being burnt the same as in the older system, but at a lower and more economical rate. It is also cleaner, reducing the problem of oil-fouling of the spark-plugs and coke formation in the cylinder and the exhaust. Almost the only motors still using premix are hand-held two-stroke devices, such as chainsaws (which must operate in any attitude) and the majority of model engines.

All two-stroke engines running on a petrol/oil mix will suffer oil starvation if forced to rotate at speed with the throttle closed, e.g. motorcycles descending long hills and perhaps when decelerating gradually from high speed by changing down through the gears. Two-stroke cars (such as those that were popular in Eastern Europe in mid-20th century) were in particular danger and were usually fitted with freewheel mechanisms in the powertrain, allowing the engine to idle when the throttle was closed, requiring the use of the brakes in all slowing situations.

Large two-stroke engines, including diesels, normally use a sump lubrication system similar to four-stroke engines. The cylinder must still be pressurised, but this is not done from the crankcase, but by an ancillary supercharger.

Two-stroke Reversibility

For the purpose of this discussion, it is convenient to think in motorcycle terms, where the exhaust pipe faces into the cooling air stream, and the crankshaft commonly spins in the same axis and direction as do the wheels i.e. "forward". Some of the considerations discussed here apply to four-stroke engines (which cannot reverse their direction of rotation without considerable modification), almost all of which spin forward, too.

Regular gasoline two-stroke engines will run backwards for short periods and under light load with little problem, and this has been used to provide a reversing facility in microcars, such as the Messerschmitt KR200, that lacked reverse gearing. Where the vehicle has electric starting, the motor will be turned off and restarted backwards by turning the key in the opposite direction. Two-stroke golf carts have used a similar kind of system. Traditional flywheel magnetos (using contact-breaker points, but no external coil) worked equally well in reverse because the cam controlling the points is symmetrical, breaking contact before TDC equally well whether running forwards or backwards. Reed-valve engines will run backwards just as well as piston-controlled porting, though rotary valve engines have asymmetrical inlet timing and will not run very well.

There are serious disadvantages to running any engine backwards under load for any length of time, and some of these reasons are general, applying equally to both two-stroke and four-stroke engines. Some of this disadvantage is intrinsic, unavoidable even in the case of a complete re-design. The problem comes about because in "forwards" running the major thrust face of the piston is on the back face of the cylinder which, in a two-stroke particularly, is the coolest and best lubricated part. The forward face of the piston is less well-suited to be the major thrust face since it covers and uncovers the exhaust port in the cylinder, the hottest part of the engine, where piston lubrication is at its most marginal. The front face of the piston is also more vulnerable since the exhaust port, the largest in the engine, is in the front wall of the cylinder. Piston skirts and rings risk being extruded into this port, so it is always better to have them pressing hardest on the back wall (where there are only the transfer ports) and there is good support. In some engines, the small end is offset to reduce thrust in the intended rotational direction and the forward face of the piston has been made thinner and lighter to compensate - but when running backwards, this weaker forward face suffers increased mechanical stress it was not designed to resist.

Large two-stroke ship diesels are sometimes made to be reversible. Like four-stroke ship engines (some of which are also reversible) they use mechanically-operated valves, so require additional camshaft mechanisms.

On top of other considerations, the oil-pump of a modern two-stroke may not work in reverse, in which case the engine will suffer oil starvation within a short time. Running a motorcycle engine backwards is relatively easy to initiate, and in rare cases, can be triggered by a back-fire. It is not advisable.

Model aeroplane engines with reed-valves can be mounted in either tractor or pusher configuration without needing to change the propeller. These motors are compression ignition, so there are no ignition timing issues and little difference between running forward and running backward.

Chapter 3

Four-stroke Engine

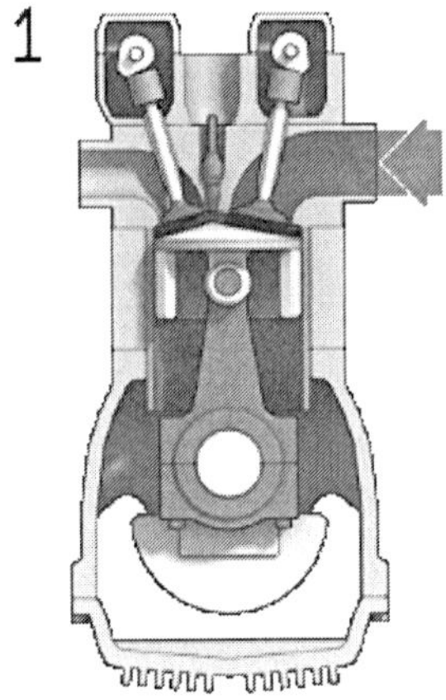

Figure 1: Four-stroke cycle used in gasoline/petrol engines. The right blue side is the intake and the left yellow side is the exhaust. The cylinder wall is a thin sleeve surrounded by cooling liquid.

A four-stroke engine, also known as four-cycle, is an internal combustion engine in which the piston completes four separate strokes—intake, compression, power, and exhaust—during two separate revolutions of the engine's crankshaft, and one single thermodynamic cycle.

There are two common types of engines, which are closely related to each other but have major differences in their design and behaviour. The earliest of these to be developed is the Otto cycle engine which was developed in 1876 by Nikolaus August Otto in Cologne, Germany, after the operation principle described by Alphonse_Beau_de_Rochas in 1861. This engine is most often referred to as a petrol engine or gasoline engine, after the fuel that powers it. The second type of four-cycle engine is the Diesel engine developed in 1893 by Rudolph Diesel, also of Germany. Diesel created his engine to maximise efficiency which was lacking in the Otto engine. There are several major differences between the Otto cycle engine and the four-cycle diesel engine. The diesel engine is made in both a two-cycle and a four-cycle version. Ironically Otto's company Deutz AG produces primarily diesel engines in the modern era.

The Otto cycle is named after the 1876 engine of Nikolaus A. Otto, who built a successful four-cycle engine which was based on the work of Jean Joseph Etienne Lenoir. It was the third engine type that Otto developed. It used a sliding flame gateway for ignition of its fuel which was a mixture of illuminating gas and air. After 1884 Otto also developed the magneto allowing the use of an electrical spark for ignition, which had been unreliable on the Lenoir engine.

Today, the internal combustion engine (ICE) is used in motorcycles, automobiles, boats, trucks, aircraft, ships, heavy duty machinery, and in its original intended use as stationary power both for kinetic and electrical power generation. Diesel engines are found in virtually all heavy duty applications such as trucks, ships, locomotives, power generation, and stationary power. Many of these diesel engine are two-cycle with power ratings up to 105,000 hp (78,000 kW).

The four cycles refer to intake, compression, combustion (power), and exhaust cycles that occur during two crankshaft rotations per power cycle of the four-cycle engines. The cycle begins at *Top Dead Centre* (TDC), when the piston is farthest away from the axis of the crankshaft. A cycle refers to the full travel of the piston from Top Dead Centre (TDC) to Bottom Dead Centre (BDC).

1. Intake stroke: on the *intake* or *induction* stroke of the piston, the piston descends from the top of the cylinder to the bottom of the cylinder, reducing the pressure inside the cylinder. A mixture of fuel and air, or just air in a diesel engine, is forced by atmospheric (or greater) pressure into the cylinder through the intake port. The intake valve(s) then close. The volume of air/fuel mixture that is drawn into the cylinder, relative to the volume of the cylinder is called, the volumetric efficiency of the engine.
2. Compression stroke: with both intake and exhaust valves closed, the piston returns to the top of the cylinder compressing the air, or fuel-air mixture into the combustion chamber of the cylinder head.
3. Power stroke: this is the start of the second revolution of the engine. While the piston is close to Top Dead Centre, the

compressed air–fuel mixture in a gasoline engine is ignited, usually by a spark plug, or fuel is injected into the diesel engine, which ignites due to the heat generated in the air during the compression stroke. The resulting massive pressure from the combustion of the compressed fuel-air mixture forces the piston back down toward bottom dead centre.

4. Exhaust stroke: during the *exhaust* stroke, the piston once again returns to top dead centre while the exhaust valve is open. This action evacuates the burnt products of combustion from the cylinder by expelling the spent fuel-air mixture out through the exhaust valve(s).

History—The Otto Cycle

Figure 2: An Otto Engine from 1920s US Manufacture

Nikolaus August Otto as a young man was a travelling salesman for a grocery concern. In his travels he encountered the internal combustion engine built in Paris by Belgian expatriate Jean Joseph Etienne Lenoir. In 1860 Lenoir succeeded in creating a double acting engine which ran on illuminating gas at 4% efficiency. The 18 litre Lenoir Engine was able to produce only 2 horsepower. The Lenoir engine ran on the illuminating gas that was made from coal which had been developed in Paris by Philip Lebon.

In testing a replica of the Lenoir engine in 1861 Otto became aware of the effects of compression on the fuel charge. In 1862 Otto attempted to produce an engine to improve on the poor efficiency and reliability of the Lenoir engine. He tried to create an engine which would compress the fuel mixture prior to ignition, but failed as that engine would run no more than a few minutes prior to its destruction. Many engineers were also trying to solve the problem with no success.

In 1864 Otto and Eugen Langen founded the first internal combustion engine production company NA Otto and Cie (NA Otto and Company). Otto and Cie succeeded in creating a successful atmospheric engine that same year.

The factory ran out of space and was moved to the town of Deutz, Germany in 1869 where the company was renamed to Deutz Gasmotoren fabrik AG (The Deutz Gas Engine Manufacturing Company). In 1872 Gottlieb Daimler was technical director and Wilhelm Maybach was the head of engine design. Daimler was a gunsmith who had also worked on the Lenoir engine previously.

By 1876 Otto and Langen succeeded in creating the first internal combustion engine that compressed the fuel mixture prior to combustion for far higher efficiency than any engine created to this time.

Gottlieb Daimler and Wilhelm Maybach left their employ at Otto and Cie and developed the first high speed Otto engine in 1883. In 1885 they produced the first automobile to be equipped with an Otto engine. The Petroleum Reitwagen used a hot tube ignition system and the fuel known as Ligroin to become the worlds first internal combustion engine power vehicle using a four-cycle engine based on Nikolaus Otto's design. The following year Karl Benz produced a four-cycle engined automobile which is called the world's first car by some.

In 1884 Otto's company, now known as Gasmotoren fabrik Deutz (GFD) developed electric ignition and the carburetor.

In 1890 Daimler and Maybach formed a company known as Daimler Motoren Gesellschaft. Today that company is known as Daimler-Benz.

The Diesel Cycle

The diesel engine is a technical refinement of the 1876 Otto Cycle engine. Where Otto had realised in 1861 that the efficiency of the engine could be increased by first compressing the fuel mixture prior to its ignition, Rudolph Diesel wanted to develop a more efficient type of engine that could run on much heavier fuel.

The Lenoir, Otto Atmospheric, and Otto Compression engines (both 1861 and 1876) were designed to run on Illuminating Gas (coal gas). With the same motivation as Otto, Diesel wanted to create an engine that would give small industrial concerns their own power source to enable them to compete against larger companies, and like

Otto to get away from the requirement to be tied to a municipal fuel supply. Like Otto, it took more than a decade to produce the high compression engine which would self ignite its fuel when that fuel sprayed into the cylinder. Diesel used an air spray combined with fuel in his first engine.

During initial development, one of the engines burst nearly killing Diesel. He persisted and finally created an engine in 1893. The high compression engine which ignites its fuel by the heat of compression is now called the Diesel engine whether it is a four-cycle or a two cycle design.

The four-cycle diesel engine has been used in the majority of heavy duty applications for many decades. Chief among the reasons for this is that it uses a heavy fuel which contains more energy, requires less refinement, and is cheaper to make (although in some areas of the world diesel fuel costs more than gasoline). The most efficient Otto Cycle engines run near 30% efficiency.

The Volkswagen Jetta TDI 1.9 Litre engine achieves 46%. It uses an advanced design with turbocharging and direct fuel injection. Some BMW ship Diesels with ceramic insulation have exceeded 60% efficiency.

Both Audi and Peugeot compete in the endurance races of the Le Mans Series with race cars having diesel engines. These are four-cycle, four-valve and high revving, turbocharged diesels which dominate largely due to fuel economy and having to make fewer stops.

Thermodynamic Analysis

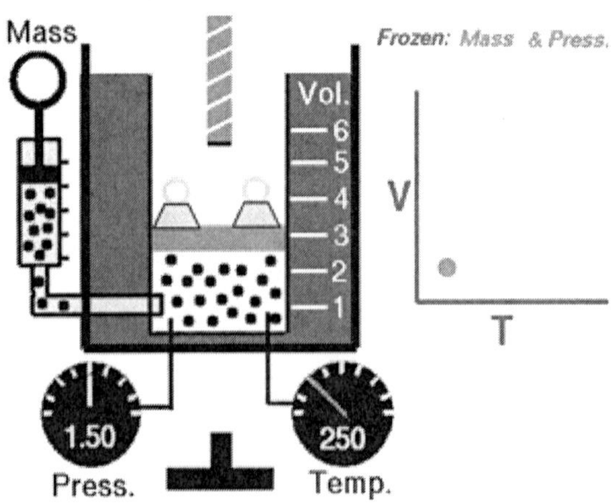

Figure 3: Charles Law

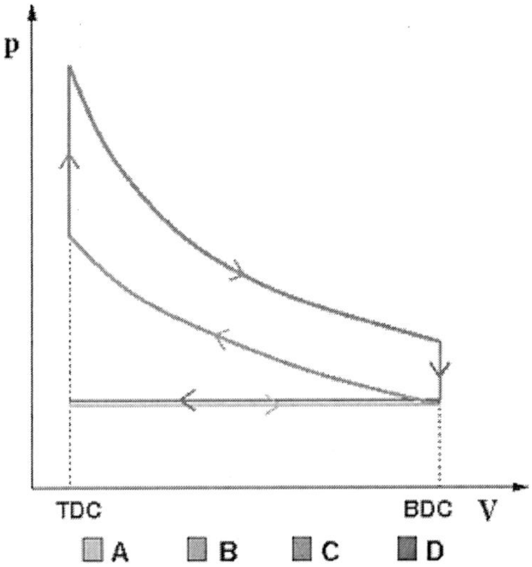

Figure 4: The idealised four-stroke Otto cycle p-V diagram: the intake (A) stroke is performed by an isobaric expansion, followed by the compression (B) stroke, performed by an adiabatic compression. Through the combustion of fuel an isochoric process is produced, followed by an adiabatic expansion, characterising the power (C) stroke. The cycle is closed by an isochoric process and an isobaric compression, characterising the exhaust (D) stroke.

The thermodynamic analysis of the actual four-stroke or two-stroke cycles is not a simple task. However, the analysis can be simplified significantly if air standard assumptions are utilised. The resulting cycle, which closely resembles the actual operating conditions, is the Otto cycle.

Octane Rating

Octane rating or octane number is a standard measure of the anti-knock properties (i.e. the performance) of a motor or aviation fuel. The higher the octane number, the more compression the fuel can withstand before detonating. In broad terms, fuels with a higher octane rating are used in high-compression engines that generally have higher performance.

Principles

Octane is a hydrocarbon liquid that is used as a reference standard to describe the tendency of gasoline, petrol, or benzin fuels to self ignite during compression prior to the desired position of the piston in the cylinder as appropriate for valve and ignition timing. The problem of premature ignition is referred to as pre-ignition and also as engine knock, which is a sound that is made when the fuel ignites too early in the compression stroke. Severe knock causes severe engine damage, such as broken connecting rods, melted pistons, melted or broken valves and other components. The octane rating is a measure of how likely a gasoline or liquid petroleum fuel is to self ignite. The higher the number, the less likely an engine is to pre-ignite and suffer damage.

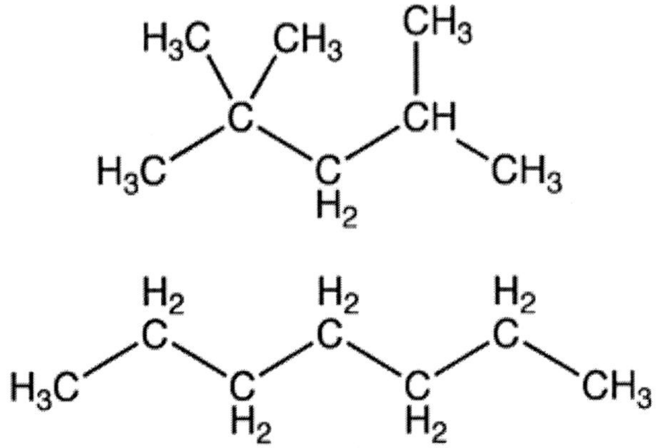

Figure 5: Isooctane (upper) has an octane rating of 100 whereas n-heptane has an octane rating of 0.

The most typically used engine management systems found in automobiles today monitor the level of knock that is being produced by the fuel being used. In modern computer controlled engines, the timing of the ignition will be automatically altered by the fuel management system to reduce the pre-ignition to an acceptable level.

The octane rating of gasoline is measured in a test engine and is defined by comparison with the mixture of 2,2,4-trimethylpentane (iso-octane) and heptane that would have the same anti-knocking capacity as the fuel under test: the percentage, by volume, of 2,2,4-trimethylpentane in that mixture is the octane number of the fuel. For example, petrol with the same knocking characteristics as a mixture of 90% iso-octane and 10% heptane would have an octane rating of 90. A rating of 90 does not mean that the petrol contains just iso-octane and heptane in these proportions, but that it has the same detonation resistance properties. Because some fuels are more knock-resistant than iso-octane, the definition has been extended to allow for octane numbers higher than 100.

Octane rating does not relate to the energy content of the fuel. It is only a measure of the fuel's tendency to burn in a controlled manner, rather than exploding in an uncontrolled manner. Where the octane number is raised by blending in ethanol, energy content per volume is reduced.

It is possible for a fuel to have a Research Octane Number (RON) greater than 100, because iso-octane is not the most knock-resistant substance available. Racing fuels, avgas, liquefied petroleum gas (LPG), and alcohol fuels such as methanol may have octane ratings of 110 or significantly higher.

Typical "octane booster" gasoline additives include MTBE, ETBE, isooctane and toluene. Lead in the form of tetra-ethyl lead was once a common additive, but since the 1970s, its use in most of the industrialised world has been restricted, and its use is currently limited mostly to aviation gasoline.

Measurement Methods

Research Octane Number (RON)

The most common type of octane rating worldwide is the Research Octane Number (RON). RON is determined by running the fuel in a test engine with a variable compression ratio under controlled conditions, and comparing the results with those for mixtures of iso-octane and n-heptane.

Motor Octane Number (MON)

There is another type of octane rating, called Motor Octane Number (MON), or the aviation lean octane rating, which is a better measure of how the fuel behaves when under load, as it is determined at 900 rpm engine speed, instead of the 600 rpm for RON. MON testing uses a similar test engine to that used in RON testing, but with a preheated fuel mixture, higher engine speed, and variable ignition timing to further stress the fuel's knock resistance.

Depending on the composition of the fuel, the MON of a modern gasoline will be about 8 to 10 points lower than the RON, however there is no direct link between RON and MON. Normally, fuel specifications require both a minimum RON and a minimum MON.

Anti-Knock Index (AKI)

In most countries, including Australia and all of those in Europe, the "headline" octane rating shown on the pump is the RON, but in Canada, the United States and some other countries, like Brazil, the headline number is the average of the RON and the MON, called the Anti-Knock Index (AKI, and often written on pumps as (R+M)/2). It may also sometimes be called the Pump Octane Number (PON).

Difference between RON and AKI

Because of the 8 to 10 point difference noted above, the octane rating shown in the United States is 4 to 5 points lower than the rating shown elsewhere in the world for the same fuel.

Observed Road Octane Number (RdON)

The final type of octane rating, called Observed Road Octane Number (RdON), is derived from testing gasolines in real world multi-cylinder engines, normally at wide open throttle. It was developed in the 1920s and is still reliable today. The original testing was done in cars on the road but as technology developed the testing was moved to chassis dynamometres with environmental controls to improve consistency.

Examples of Octane Ratings

The RON/MON values of n-heptane and iso-octane are exactly 0 and 100, respectively, by the definition of octane rating. The following table lists octane ratings for various other fuels.

Fuel	*RON*	*MON*	*AKI*
hexadecane	< -30		
n-octane	-10		
n-heptane (RON and MON 0 by definition)	0	0	0
diesel fuel	15–25		
2-methylheptane	23	23.8	
n-hexane	25	26.0	26
1-pentene	34		
2-methylhexane	44	46.4	
3-methylhexane		55.0	
1-heptene	60		
n-pentane	62	61.9	
requirement for a typical two-stroke outboard engine	69	65	67
Pertamina "Premium" gasoline in Indonesia	88		
n-butanol	92	71	83
2,2-dimethylpropane		80.2	
"regular" gasoline in Australia, New Zealand, Canada and the US	91–92	82–83	87
Pertamina "Pertamax" gasoline in Indonesia	92		
Shell "Super" in Indonesia	92		

n-butane	94	90.1	
2-methylbutane		90.3	
Pertamina "Pertamax Plus" gasoline in Indonesia	95		
Shell "Super Extra" in Indonesia	95		
Shell "FuelSave " in Malaysia	95		
"EuroSuper" or "EuroPremium" or "Regular unleaded" in Europe, "SP95" in France	95	85–86	90–91
Shell "V-Power 97" in Malaysia	97		
"SuperPlus" in Germany, Great Britain, Slovenia and Spain, "SP98" in France	98	89–90	93–94
Shell V-Power in Italy and Germany	100		
Eni(or Agip) Blu Super +(or Tech) in Italy	100		
IP Plus 100 in Italy	100		
Tamoil WR 100 in Italy	100		
San Marco Petroli F-101 in Italy (northern Italy only, just a few gas stations)	101		
Petro-Canada "Ultra 94" in Canada	101.5	88	94
Aral Ultimate 102 in Germany	102		
IES 98 Plus in Italy	98		
2,2-dimethylbutane		93.4	
2,3-dimethylbutane		94.4	
ExxonMobil Avgas 100		99.5 (min)	
Shell "V-Power Racing" in Australia - discontinued July 2008	100		
"isooctane" (RON and MON 100 by definition)	100	100	100
benzene	101		
i-butane	102	97.6	
"BP Ultimate 102 - now discontinued"	102	93–94	97–98
t-butanol	103	91	97
2,3,3-trimethylpentane	106.1	99.4	103
ethane	108		
2,2,3-trimethylpentane	109.6	99.9	105
toluene	111	95	103
E85 gasoline			94-96
propane	112	97	
2,2,3-trimethylbutane	112.1	101.3	106
xylene	117		

isopropanol	118	98	108
methanol	108.7	88.6	98.65
ethanol	108.6	89.7	99.15
2,5-Dimethylfuran	119		
methane	120	120	120
hydrogen*	> 130	very low	

*Hydrogen does not fit well into the normal definitions of octane number. It has a very high RON and a low MON, so that it has low knock resistance in practice, due to its low ignition energy (primarily due to its low dissociation energy) and extremely high flame speed. These traits are highly desirable in rocket engines, but undesirable in Otto-cycle engines. However, as a minor blending component (e.g. in a bi-fuel vehicle), hydrogen raises overall knock resistance. Flame speed is limited by the rest of the component species; hydrogen may reduce knock because of its high thermal conductivity.

Effects of Octane Rating

Higher octane ratings correlate to higher activation energies: This being the amount of applied energy required to initiate combustion. Since higher octane fuels have higher activation energy requirements, it is less likely that a given compression will cause uncontrolled ignition, otherwise known as autoignition or detonation.

It might seem odd that fuels with higher octane ratings are used in more powerful engines, since such fuels ignite less easily. However, detonation is undesirable in a spark ignition engine, and is signified by audible "pinging" or in more extreme cases "knock".

A fuel with a higher octane rating can be burnt in an engine with a high compression ratio without causing detonation, as such fuels are less prone to detonation. Compression is directly related to power and to thermodynamic efficiency, so engines that require a higher octane fuel usually develop more motive power and therefore do more work in relation to the calorific value of the fuel (BTU) being used. Power output is a function of the properties of the fuel used, as well as the design of the engine itself, and is related to octane rating of the fuel. Power is limited by the maximum amount of fuel-air mixture that can be brought into the combustion chamber. When the throttle is partly open, only a small fraction of the total available power is produced because the manifold is operating at pressures far below that of the external atmosphere (depression). In this case, the octane requirement is far lower than when the throttle is opened fully and the manifold pressure increases to almost that of the external atmosphere, or higher in the case of forced induction engines.

Many high-performance engines are designed to operate with a high maximum compression, and thus demand fuels of higher octane. A common misconception is that power output or fuel efficiency can be improved by burning fuel of higher octane than that specified by the engine manufacturer. The power output of an engine depends in part on the energy density of the fuel being burnt. Fuels of different octane ratings may have similar densities, but because switching to a higher octane fuel does not add more hydrocarbon content or oxygen, the engine cannot develop more power.

However, burning fuel with a lower octane rating than that for which the engine is designed often results in a reduction of power output and efficiency. Many modern engines are equipped with a knock sensor (a small piezoelectric microphone), which sends a signal to the engine control unit, which in turn retards the ignition timing when detonation is detected.

Retarding the ignition timing reduces the tendency of the fuel-air mixture to detonate, but also reduces power output and fuel efficiency.

Because of this, under conditions of high load and high temperature, a given engine may have a more consistent power output with a higher octane fuel, as such fuels are less prone to detonation. Some modern high performance engines are actually optimised for higher than pump premium (93 AKI in the US). The 2001 - 2007 BMW M3 with the S54 engine is one such car. Car and Driver magazine tested a car using a dynamometre, and found that the power output increased as the AKI was increased up to approximately 96 AKI.

Most fuel filling stations have two storage tanks (even those offering 3 or 4 octane levels): those motorists who purchase intermediate grade fuels are given a mixture of higher and lower octane fuels. "Premium" grade is fuel of higher octane, and the minimum grade sold is fuel of lower octane. Purchasing 91 octane fuel (where offered) simply means that more fuel of higher octane is blended with commensurately less fuel of lower octane, than when purchasing a lower grade. The detergents and other additives in the fuel are often, but not always, identical.

The octane rating was developed by chemist Russell Marker at the Ethyl Corporation in 1926. The selection of *n*-heptane as the zero point of the scale was due to its availability in high purity. Other isomers of heptane produced from crude oil have greatly different ratings.

Regional Variations

The selection of octane ratings available at the pump can vary greatly from region to region.

- Australia: "regular" unleaded fuel is 91 RON, "premium" unleaded with 95 RON is widely available, and 98 RON fuel is also reasonably common. Shell used to sell 100 RON petrol (5% ethanol content) from a small number of service stations, most of which are located in major cities (stopped in August 2008). United Petroleum sells 100 RON unleaded fuel (10% ethanol content) at a small number of its service stations (originally only two, but it has now expanded to 19 outlets).
- Bahrain: 90 and 95 (RON), standard in all petrol station in the country and advertised as (Jayyid) for Regular or 90 and (Mumtaz) for Premium or 95.
- China: 93 and 97 (RON) are commonly offered. In limited areas higher rating such as 99 RON is available. In some rural areas it can be difficult to find fuel with over 93 RON.
- Chile: 93, 95 and 97 RON are standard at almost all gas stations thorough Chile. The three types are unleaded
- Egypt: 80 RON is commonly used for all taxis and old cars and is the predominant rating in rural areas. 90 RON and 92 RON are available at almost all gas stations with a negligible price difference between them. 95 RON is becoming more common especially in the big cities and upscale suburbs. All fuels are unleaded.
- Finland: 95 and 98 (RON), advertised as such, at almost all gas stations. Most cars run on 95, but 98 is available for vehicles that need higher octane fuel. Shell offers V-Power, advertised as "over 99 octane", instead of 98. In the beginning of 2011 95 RON was replaced by 95E10, and 98 RON will be advertised as 98E5. ST1 also offers RE85 on some stations, which is E85 made from waste from which the advertised name "ReFuel" comes from.
- Germany: "Super E10" 95 RON and "Super Plus E5" 98 RON are available practically everywhere. Big suppliers like Shell or Aral offer 100 RON gasoline (Shell V-Power, Aral Ultimate) at almost every fuel station. "Normal" 91 RON is only rarely being offered, because lower production amounts make it more

expensive than "Super" 95 RON, so it is often not offered any more. Due to a new European Union law, gas stations are being required to offer and sell a minimum rate of the new mixture of "Super" 95 RON with up to 10% Ethanol branded as "Super E10" since early 2011. Due to that process the big suppliers are discontinuing the average "Super E5" 95 RON with <5% Ethanol so cars without the capability for using E10 need to use 98 RON petrol instead.

- Hong Kong: only 98 RON and 99 RON are available in the market. There have been calls to re-introduce 95 RON, but the calls have been rejected by all petrol station chains, citing that 95 RON was phased out because of market forces.
- India: India's Ordinary And Premium Petrols are of 89–91 RON. The premium petrols are generally ordinary fuels with additives, that do not really change the octane value. Two variants, "Speed 93" and "Speed 97", were launched, with RON values of 93 and 97. India's economy-class vehicles usually have compression ratios under 10:1, thus enabling them to use lower-quality petrol without engine knocking.
- Indonesia: Indonesia's "Premium" petrol rated at 88 RON and being subsidised it cost only about US$0.50/litre. Other options are "Pertamax" rated at 92 RON and the "Pertamax Plus" rated at RON 95, which is the highest octane available for automotive gasoline in Indonesia.
- Ireland: 95 RON "unleaded" is the only petrol type available through stations, although E5 (99 RON) is becoming more commonplace.
- Italy: 95 RON is the only compulsory gasoline offered (*verde*, "green"), only a few fuel stations (Agip, IP, IES, OMV) offer 98 RON as the premium type, many Shell and Tamoil stations close to the cities offer also V-Power Gasoline rated at 100 RON. Recently Agip introduced "Blu Super+", a 100 RON gasoline.
- Israel: 95 RON & 98 RON are normally available at most petrol stations. 96 RON is also available at a large number of gas stations but 95 RON is more preferred because it's cheaper and performance differences aren't very wide and noticeable. "Regular" fuel is 95 RON. All variants are unleaded.
- Latvia: 95 RON and 98 RON widely available.

- Lebanon: 95 RON and 98 RON are widely available.
- Malaysia: Had RON 92 until 2009. Replaced with "regular" unleaded fuel Ron 95 RON, "premium" fuel is rated at 97 RON (but for Shell 97 RON is V-Power, and Shell's V-Power Racing is rated at 97 RON.)
- Montenegro: 95 RON is sold as a "regular" fuel. As a "premium" fuel, 98 RON is sold. Both variants are unleaded.
- Netherlands: 95 RON "Euro" and 98 RON "Super" are sold at practically every station. Shell V-Power is a 97 RON (labelled as 95 due to the legalities of only using 95 or 98 labelling), some independent test have shown that one year after introduction it was downgraded to 95 RON, whereas in neighbouring Germany Shell V-Power consists of the regular 100 RON fuel.
- New Zealand: 91 RON "Regular" and 95 RON "Premium" are both widely available. 98 RON is available instead of 95 RON at some service stations in larger urban areas.
- Philippines: A brand of Petron, Petron Blaze is rated at 100 RON (the only brand of gasoline in the Philippines without an ethanol blend). Other "super premium" brands like Petron XCS, Calex Gold, Shell V-Power are rated at 95-97 RON, while Petron Xtra Unleaded, Caltex Silver, and Shell Super Unleaded are rated at 93 RON.
- Poland: Eurosuper 95 (RON 95) is sold in every gas station. Super Plus 98 (RON 98) is available in most stations, sometimes under brand (Orlen - Verva, BP - Ultimate, Shell - V-Power) and usually containing additives. Shell offers V-Power Racing fuel which is rated RON 100.
- Russia and CIS countries: 80 RON (76 MON) is the minimum available, the standard is 92 RON and 95 RON. 98 RON is available on some stations but it's usually quite expensive compared to the lower octane rating fuels.
- Saudi Arabia: Two types of fuel are available at all gas stations in Saudi Arabia. "Premium 91" (RON 91) where the pumps are coloured green, and "Super Premium 95" (RON 95) where the pumps are coloured red. While gas stations in Saudi Arabia are privatised, the prices are regulated by the authorities and have a fixed at S.A.R. 0.45 (U.S. $0.12) and S.A.R. 0.60 (U.S. $0.16) per litre respectively. Prior to 2006, only Super Premium

RON 95 was available and the pumps weren't coloured in any specific order. The public didn't know what Octane rating was, therefore big educating campaigns were spread, telling the people to use the "red gas" only for high end cars, and save money on using the "green gas" for regular cars and trucks.

- South Africa: "regular" unleaded fuel is 95 RON in coastal areas with most fuel stations optionally offering 97 RON. Inland (higher elevation) "regular" unleaded fuel is 93 RON; once again most fuel stations optionally offer 95 RON.
- Spain: 95 RON "Euro" is sold in every station with 98 RON "Super" being offered in most stations. Many stations around cities and highways offer other high-octane "premium" brands.
- Sri Lanka: In Ceypetco filling stations, 90 RON is the regular petrol and 95 RON is called 'Super Petrol', which comes at a premium price. In LIOC filling stations, 90 RON remains as regular petrol and 92 RON is available as 'Premium Petrol'. The cost of premium petrol is lower than the cost of super petrol.
- Taiwan: 92 RON, 95 RON and 98 RON are widely available at gas stations in Taiwan.
- Thailand: 91 RON and 95 RON are widely available.
- Turkey: 95 RON and 98 RON are widely available in gas stations. 92 RON (Regular) has been dropped in 2006.
- Ukraine: the standard gasoline is 95 RON, but 92 RON gasoline is also widely available and popular as a less expensive replacement for 95 RON gasoline. 80 RON gasoline is available for old cars and motorcycles.
- United Kingdom: 'regular' petrol has an octane rating of 95 RON, with 97 RON fuel being widely available as the *Super Unleaded.* Tesco and Shell both offer 99 RON fuel. In April 2006, BP started a public trial of the super-high octane petrol *BP Ultimate Unleaded 102*, which as the name suggests, has an octane rating of 102 RON. Although BP Ultimate Unleaded (with an octane rating of 97 RON) and BP Ultimate Diesel are both widely available throughout the UK, BP Ultimate Unleaded 102 was available throughout the UK in only 10 filling stations, and was priced at about two and half times more than their 97 RON fuel. In March 2010, BP stopped sales

of Ultimate Unleaded 102, citing the closure of their speciality fuels manufacturing facility. Shell V-Power is also available, but in a 99 RON octane rating, and Tesco fuel stations also supply the Greenergy produced 99 RON "Tesco 99".

- United States: in the US octane rating is displayed in AKI. In the Rocky Mountain (high elevation) states, 85 AKI (90 RON) is the minimum octane, and 91 AKI (95 RON) is the maximum octane available in fuel. The reason for this is that in higher-elevation areas, a typical naturally aspirated engine draws in less air mass per cycle because of the reduced density of the atmosphere. This directly translates to less fuel and reduced absolute compression in the cylinder, therefore deterring knock. It is safe to fill a carbureted car that normally takes 87 AKI fuel at sea level with 85 AKI fuel in the mountains, but at sea level the fuel may cause damage to the engine. A disadvantage to this strategy is that most turbocharged vehicles are unable to produce full power, even when using the "premium" 91 AKI fuel. In some east coast states, up to 94 AKI (98 RON) is available. In Colorado as well as parts of the Midwest (primarily Minnesota, Iowa, Illinois and Missouri) ethanol-based E-85 fuel with 105 AKI is available. Often, filling stations near US racing tracks will offer higher octane levels such as 100 AKI. California fuel stations will offer 87, 89, and 91 AKI (91, 93 and 95 RON) octane fuels, and at some stations, 100 AKI or higher octane, sold as racing fuel.
- Venezuela: 91 RON and 95 RON gasoline is available nationwide, in all PDV gas stations. 95 RON petrol is the most widely used in the country, although most cars in Venezuela would work with 91 RON gasoline. This is because petrol prices are heavily subsided by the government. All gasoline in Venezuela is unleaded.
- Vietnam: 92 is in every gas station and 95 is in the urban areas.

Otto Engines

During the compression cycle of a compressed charge internal combustion engine the temperature of the fuel air mixture rises as described by Charles Law solely due to the compression of the gases. The temperature rise is several hundreds of degrees.

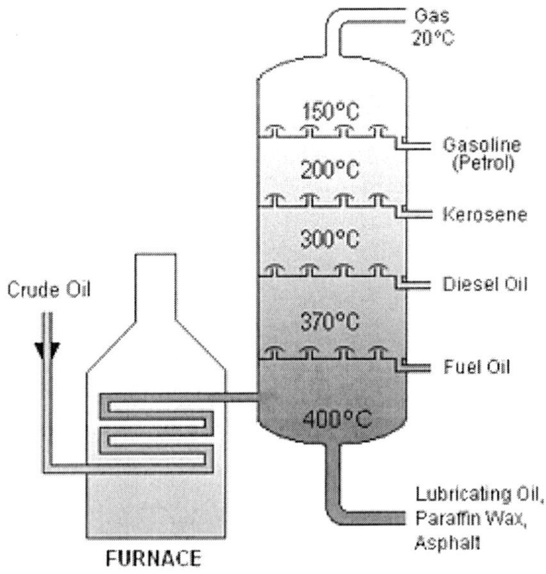

Figure 6: A Refractory Tower showing the differing weights of various products.

The fuels used in four-cycle engines are most typically fractions of crude oil, coal tar, oil shale, or sands which are produced in a process called Petroleum Cracking. The ignition temperature of the fuel that is refracted is related to its weight. It is separated by being heating and is extracted at different heights in the refractory tower. The higher the fuel vapour rises in the tower the lower its weight and the less energy it contains. In refracting petroleum, there is a standard weight of fuels and products that is withdraw and which is associated with a specific extracted material. Gasoline is a light refractory product and is called a light fraction. As a light fraction it has a relatively low flash point (that is the temperature at which it starts to burn when mixed with an oxidiser).

A fuel with a low flash point may self ignite during compression, and can also be ignited by carbon deposits left in the cylinder or head of a dirty engine. In an internal combustion engine self ignition can occur at unexpected times. During the normal operation of the engine as the fuel mixture is being compressed an electric arc is created to ignite the fuel. At low rpm this occurs close to TDC (Top Dead Centre). As engine rpm rises the spark point is moved forward so that the fuel charge can be ignited at a more efficient point in fuel charge compression to allow the fuel to start burning even while it is still being compressed. This produces more effective power based on the rising molecular density of the working medium, since this is the essence of efficiency in the compressed charge IE engine. A denser working medium (the air fuel mixture) will experience a greater heat, and therefore pressure, rise on less few when its molecules are more densely packed together.

We can see this in two of the designs of the Otto engines. The non-compression engine operated at 12% efficiency. The compressed charge engine had an operating efficiency of 30%. A Diesel engine can reach as high as 70% (Diesel's lab engine tested at 75.6% efficiency, VW TDI is at 46%).

The problem with compressed charge engines is that the temperature rise of the compressed charge can cause pre-ignition. If this occurs at the wrong time and is too energetic, it can destroy the engine. Fractions of petroleum have widely varying flash points (the temperatures at which the fuel may self ignite). This must be taken into account in engine and fuel design.

In engines, the spark is retarded when the engine is being started, and progresses only to an appropriate amount based on engine rpm. This is determined by laboratory research. As the engine revolves faster it can accept earlier ignition since the moving flame front will not have time to be destructive.

In fuel, the tendency for the compressed fuel mixture to ignite early is limited by the chemical composition of the fuel. There are several grades of fuel to accommodate differing performance levels of engines. The fuel is altered to change its self ignition temperature. There are several ways to do this. As engines are designed with higher compression ratios the result is that pre-ignition is much more likely to occur since the fuel mixture will be compressed to a higher temperature prior to deliberate ignition. The higher temperature will more effectively evapourate fuels such as gasoline and is factor in a higher compression engine being higher efficiency. Higher Compression ratios also mean that the distance that the piston can push to produce power is greater (which is called the Expansion ratio).

So there must be a standardised test and a standard reference system to describe how likely a fuel is to self ignite. The octane rating is a measure of the fuel's resistance to self-ignition. A fuel with a higher numerical octane rating allows for a much higher compression ratio, which extracts more energy from the fuel and more effectively converts that energy into useful work while at the same time preventing engine damage from pre-ignition. High Octane fuel is also more expensive.

Diesel Engines

Diesel engines by their nature do not have concerns with pre-ignition. They have a concern with whether or not combustion can be started. The description of how likely Diesel fuel is to ignite is called the Cetane rating. Because Diesel fuels are of low volatility, they can be very hard to start when cold. Various techniques are used to start a cold Diesel engine, the most common being the use of a glow plug. In some applications, such as in burning used cooking oil, the fuel itself is solid and must be heated to liquify prior to use. A common complaint here is that the exhaust may have the odour of French Fries.

Design and Engineering Principles—Power Output Limitations

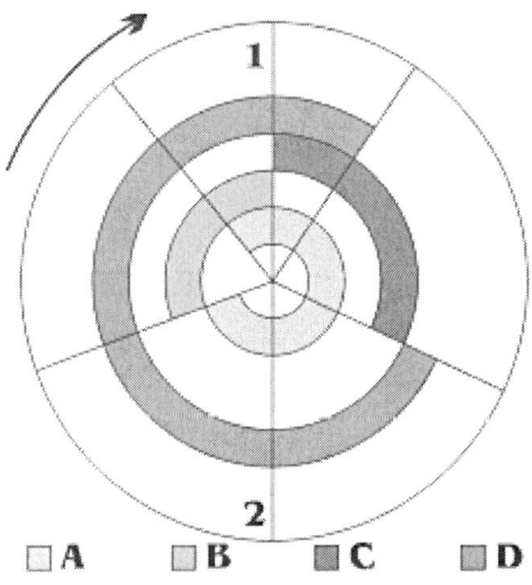

Figure 7: The four-stroke cycle

1=TDC

2=BDC

A: Intake

B: Compression

C: Power

D: Exhaust

The maximum amount of power generated by an engine is determined by the maximum amount of air ingested. The amount of power generated by a piston engine is related to its size (cylinder volume), whether it is a two-stroke or four-stroke design, volumetric efficiency, losses, air-to-fuel ratio, the calorific value of the fuel, oxygen content of the air and speed (RPM). The speed is ultimately limited by material strength and lubrication. Valves, pistons and connecting rods suffer severe acceleration forces. At high engine speed, physical breakage and piston ring flutter can occur, resulting in power loss or even engine destruction. Piston ring flutter occurs when the rings oscillate vertically within the piston grooves they reside in. Ring flutter compromises the seal between the ring and the cylinder wall which results in a loss of cylinder pressure and power. If an engine spins too quickly, valve springs cannot act quickly enough to close the valves. This is commonly referred to as 'valve float', and it can result in piston to valve contact, severely damaging the engine. At high speeds the lubrication of piston cylinder wall interface tends to break down. This limits the piston speed for industrial engines to about 10 m/s.

Intake/Exhaust Port Flow

The output power of an engine is dependent on the ability of intake (air–fuel mixture) and exhaust matter to move quickly through valve ports, typically located in the cylinder head. To increase an engine's output power, irregularities in the intake and exhaust paths, such as casting flaws, can be removed, and, with the aid of an air flow bench, the radii of valve port turns and valve seat configuration can be modified to reduce resistance. This process is called porting, and it can be done by hand or with a CNC machine.

Supercharging

One way to increase engine power is to force more air into the cylinder so that more power can be produced from each power stroke. This can be done using some type of air compression device known as a supercharger, which can be powered by the engine crankshaft.

Supercharging increases the power output limits of an internal combustion engine relative to its displacement. Most commonly, the supercharger is always running, but there have been designs that allow it to be cut out or run at varying speeds (relative to engine speed). Mechanically driven supercharging has the disadvantage that some of the output power is used to drive the supercharger, while power is wasted in the high pressure exhaust, as the air has been compressed twice and then gains more potential volume in the combustion but it is only expanded in one stage.

Turbocharging

A turbocharger is a supercharger that is driven by the engine's exhaust gases, by means of a turbine. It consists of a two piece, high-speed turbine assembly with one side that compresses the intake air, and the other side that is powered by the exhaust gas outflow.

When idling, and at low-to-moderate speeds, the turbine produces little power from the small exhaust volume, the turbocharger has little effect and the engine operates nearly in a naturally aspirated manner. When much more power output is required, the engine speed and throttle opening are increased until the exhaust gases are sufficient to 'spin up' the turbocharger's turbine to start compressing much more air than normal into the intake manifold.

Turbocharging allows for more efficient engine operation because it is driven by exhaust pressure that would otherwise be (mostly) wasted, but there is a design limitation known as turbo lag.

The increased engine power is not immediately available due to the need to sharply increase engine RPM, to build up pressure and to spin up the turbo, before the turbo starts to do any useful air compression. The increased intake volume causes increased exhaust and spins the turbo faster, and so forth until steady high power operation is reached. Another difficulty is that the higher exhaust pressure causes the exhaust gas to transfer more of its heat to the mechanical parts of the engine.

Rod and Piston-to-stroke Ratio

The rod-to-stroke ratio is the ratio of the length of the connecting rod to the length of the piston stroke. A longer rod will reduce the sidewise pressure of the piston on the cylinder wall and the stress forces, hence increasing engine life. It also increases the cost and engine height and weight.

A "square engine" is an engine with a bore diametre equal to its stroke length. An engine where the bore diametre is larger than its stroke length is an oversquare engine, conversely, an engine with a bore diametre that is smaller than its stroke length is an undersquare engine.

Valvetrain

The valves are typically operated by a camshaft rotating at half the speed of the crankshaft. It has a series of cams along its length, each designed to open a valve during the appropriate part of an intake or exhaust stroke. A tappet between valve and cam is a contact surface on which the cam slides to open the valve. Many engines use one or more camshafts "above" a row (or each row) of cylinders, as in the illustration, in which each cam directly actuates a valve through a flat tappet. In other engine designs the camshaft is in the crankcase, in which case each cam contacts a push rod, which contacts a rocker arm which opens a valve. The overhead cam design typically allows higher engine speeds because it provides the most direct path between cam and valve.

Valve Clearance

Valve clearance refers to the small gap between a valve lifter and a valve stem that ensures that the valve completely closes. On engines with mechanical valve adjustment excessive clearance will cause noise from the valve train. Typically the clearance has to be readjusted each 20,000 miles (32,000 km) with a feeler gauge.

Most modern production engines use hydraulic lifters to automatically compensate for valve train component wear. Dirty engine oil may cause lifter failure.

Energy Balance

Otto engines are about 30% efficient; in other words, 30% of the energy generated by combustion is converted into useful rotational energy at the output shaft of the engine, while the remainder being losses due to friction, engine accessories, and waste heat. There are a number of ways to recover some of the energy lost to waste heat. The use of a Turbocharger in Diesel engines is very effective by boosting incoming air pressure and in effect provides the same increase in performance as having more displacement.

The Mack Truck company decades ago developed a turbine system which converted waste heat into kinetic energy which was fed back into the engine's transmission. In 2005, BMW announced the development of the turbosteamer, a two stage heat recovery system similar to the Mack system that recovers 80% of the energy in the exhaust gas and raised the efficiency of the Otto engines it is applied to by 15%.

By contrast, a six-stroke engine may convert more than 50% of the energy of combustion into useful rotational energy.

Modern engines are often intentionally built to be slightly less efficient than they could otherwise be. This is necessary for emission controls such as exhaust gas recirculation and catalytic converters that reduce smog and other atmospheric pollutants. Reductions in efficiency may be counteracted with an engine control unit using lean burn techniques.

In the United States, the Corporate Average Fuel Economy will mandate that vehicles must achieve an average of 35.5 miles per gallon (mpg) compared to the current standard of 25 mpg. As automakers look to meet these standards by 2016, new ways of engineering the traditional internal combustion engine (ICE) could have to be considered.

Some potential solutions to increase fuel efficiency to meet new mandates include firing after the piston is farthest from the crankshaft, known as top dead centre, and applying the Miller cycle. Together, this redesign could significantly reduce fuel consumption and NOx emissions.

Chapter 4

Six-stroke Engine

The six-stroke engine is a type of internal combustion engine based on the four-stroke engine, but with additional complexity intended to make it more efficient and reduce emissions. Two types of six-stroke engine have been developed since the 1990s:

In the first approach, the engine captures the heat lost from the four-stroke Otto cycle or Diesel cycle and uses it to power an additional power and exhaust stroke of the piston in the same cylinder. Designs use either steam or air as the working fluid for the additional power stroke. The pistons in this type of six-stroke engine go up and down three times for each injection of fuel. There are two power strokes: one with fuel, the other with steam or air. The currently notable designs in this class are the Crower six-stroke engine, invented by Bruce Crower of the U.S. ; the Bajulaz engine by the Bajulaz S.A. company of Switzerland; and the Velozeta Six-stroke engine built by the College of Engineering, at Trivandrum in India.

The second approach to the six-stroke engine uses a second opposed piston in each cylinder that moves at half the cyclical rate of the main piston, thus giving six piston movements per cycle. Functionally, the second piston replaces the valve mechanism of a conventional engine but also increases the compression ratio. The currently notable designs in this class include two designs developed independently: the Beare Head engine, invented by Australian Malcolm Beare, and the German Charge pump, invented by Helmut Kottmann.

Engine Types

Griffin Six-stroke Engine

In 1883, the Bath-based engineer Samuel Griffin was an established maker of steam and gas engines. He wished to produce an internal combustion engine, but without paying the licensing costs of the Otto patents. His solution was to develop a 'Patent slide valve' and a single-acting six-stroke engine using it. By 1886, Scottish steam locomotive maker Dick, Kerr & Co. saw a future in large oil engines and licensed the Griffin patents. These were double acting, tandem engines and sold under the name "Kilmarnock". A major market for the Griffin engine was in electricity generation, where they developed a reputation for happily running light for long periods, then suddenly being able to take up a large demand for power. Their large heavy construction didn't suit them to mobile use, but they were capable of burning heavier and cheaper grades of oil.

The key principle of the "Griffin Simplex" was a heated exhaust-jacketed external vapouriser, into which the fuel was sprayed. The temperature was held around 550 °F (288 °C), sufficient to physically vapourise the oil but not to break it down chemically. This fractional distillation supported the use of heavy oil fuels, the unusable tars and asphalts separating out in the vapouriser.

Hot bulb ignition was used, which Griffin termed the *'Catathermic Igniter'*, a small isolated cavity connected to the combustion chamber. The spray injector had an adjustable inner nozzle for the air supply, surrounded by an annular casing for the oil, both oil and air entering at 20 lbs sq in. pressure, and being regulated by a governor.

Griffin went out of business in 1923.

Only two known examples of a Griffin six-stroke engine survive. One is in the Anson engine museum. The other was built in 1885 and for some years was in the Birmingham Museum of Science and Technology, but in 2007 it returned to Bath and the Museum of Bath at Work.

Bajulaz Six-stroke Engine

The Bajulaz six-stroke engine is similar to a regular combustion engine in design. There are however modifications to the cylinder head, with two supplementary fixed capacity chambers: a combustion chamber and an air preheating chamber above each cylinder. The combustion chamber receives a charge of heated air from the cylinder; the injection of fuel begins an isochoric burn which increases the thermal efficiency compared to a burn in the cylinder. The high pressure achieved is then released into the cylinder to work the power or expansion stroke. Meanwhile a second chamber which blankets the combustion chamber, has its air content heated to a high degree by heat passing through the cylinder wall. This heated and pressurised air is then used to power an additional stroke of the piston.

The claimed advantages of the engine include reduction in fuel consumption by at least 40%, two expansion strokes in six strokes, multi-fuel usage capability, and a dramatic reduction in pollution.

The Bajulaz Six-Stroke Engine was invented in 1989 by the Bajulaz S A company, based in Geneva, Switzerland; it has U.S. Patent 4,809,511 and U.S. Patent 4,513,568.

The Bajulaz six-stroke engine features:

- Reduction in fuel consumption by at least 40%
- Two expansion (work) strokes in six strokes
- Multifuel, including liquefied petroleum gas
- Dramatic reduction in air pollution
- Costs comparable to those of a four-stroke engine.

Velozeta Six-stroke Engine

In a Velozeta engine, during the exhaust stroke, fresh air is injected into the cylinder, which expands by heat and therefore forces the piston down for an additional stroke. The valve overlaps have been removed and the two additional strokes using air injection provide for better gas scavenging. The engine seems to show 40% reduction in fuel consumption and dramatic reduction in air pollution. Its specific power is not much less than that of a four-stroke Gasoline engine. The engine can run on a variety of fuels, ranging from Gasoline and diesel to LPG. An altered engine shows a 65% reduction in carbon monoxide pollution when compared with the four stroke engine from which it was developed.

The engine was developed in 2006 by a team of mechanical engineering students (Aaron Joseph George, Arun K Nair, Boby Sebastian, Krishnaraj U) of the College of Engineering, Trivandrum.

Crower Six-stroke Engine

In a six-stroke engine prototyped in the U.S. by Bruce Crower, water is injected into the cylinder after the exhaust stroke, and is instantly turned to steam, which expands and forces the piston down for an additional power stroke. Thus, waste heat that most engines require an air or water cooling system to discharge is captured and put to use driving the piston. Crower estimated that his design would reduce fuel consumption by 40% by generating the same power output at a lower RPM. The weight associated with a cooling system could be eliminated, but that would be balanced by a need for a water tank in addition to the normal fuel tank.

The Crower six-stroke engine was an experimental design that attracted media attention in 2006 because of an interview given by 75 year old American inventor Pankaj Sharma of Guru Nanak Dev University, Punjab, India has applied for a patent on his design. That patent application was subsequently abandoned. Leonard Dyer invented the first six-stroke internal combustion water injection engine in 1915, which is very similar to Crower's design. A dozen more similar patents have been issued since. Dyer's six-stroke engine features:

- No cooling system required
- Improves a typical engine's fuel consumption
- Requires a supply of pure water to act as the medium for the second power stroke.

Beare Head

The term "Six Stroke" was coined by the inventor of the Beare Head, Malcolm Beare. The technology combines a four stroke engine bottom end with an opposed piston in the cylinder head working at half the cyclical rate of the bottom piston. Functionally, the second piston replaces the valve mechanism of a conventional engine first use of "sixstroke" 1994

M4+2

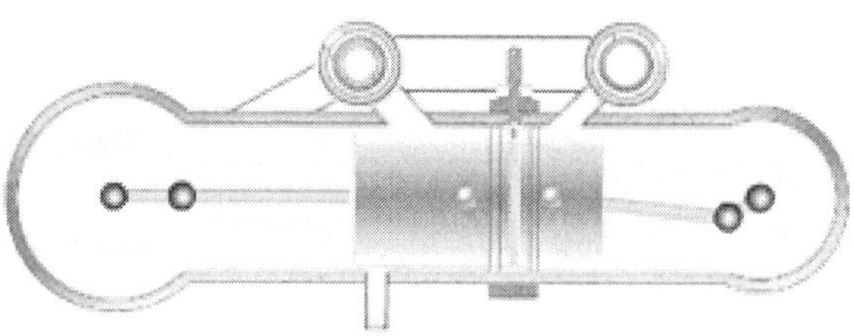

Figure 1: The M4+2 engine working cycle animation

The M4+2 engines have much in common with the Beare Head engines, combining two opposed pistons in the same cylinder. One piston working at half the cyclical rate of the other. But while the main function of the second piston in a Beare Head engine is to replace the valve mechanism of a conventional four stroke engine, the M4+2 take the principle one step further.

Piston Charger Engine

In this engine, similar in design to the Beare head, a "piston charger" replaces the valve system. The piston charger charges the main cylinder and simultaneously regulates the inlet and the outlet aperture leading to no loss of air and fuel in the exhaust. In the main cylinder, combustion takes place every turn as in a two-stroke engine and lubrication as in a four-stroke engine. Fuel injection can take place in the piston charger, in the gas transfer channel or in the combustion chamber. It is also possible to charge two working cylinders with one piston charger.

The combination of compact design for the combustion chamber together with no loss of air and fuel is claimed to give the engine more torque, more power and better fuel consumption. The benefit of less moving parts and design is claimed to lead to lower manufacturing costs. Good for hybrid technology and stationary engines. The engine is claimed to be suited to alternative fuels since there is no corrosion or deposits left on valves. The six strokes are: aspiration, precompression, gas transfer, compression, ignition and ejection. This is an invention of Helmut Kottmann from Germany, working 25 years at MAHLE GmbH piston and cylinder construction.

Related U.S. Patents

- 1217788 Internal combustion and steam engine Feb 27, 1917. Hugo F. Liedtke seems to be one of the first to contemplate alternating between internal combustion and steam injection into the combustion chamber.
- 1339176 Internal combustion engine May 4, 1920. Leonard H. Dyer invented the first 6-stroke internal combustion/water-injection engine in 1915.
- 3964263 Six cycle combustion and fluid vapourisation engine Jun 22, 1976
- 4143518 Internal combustion and steam engine Mar 13, 1979
- 4301655 Combination internal combustion and steam engine Nov 24, 1981
- 4433548 Combination internal combustion and steam engine Feb 28, 1984
- 4489558 Compound internal combustion engine and method for its use Dec 25, 1984
- 4489560 Compound internal combustion engine and method for its use Dec 25, 1984
- 4736715 Engine with a six-stroke cycle, variable compression ratio, and constant stroke Apr 12, 1988

- 4917054 Six-stroke internal combustion engine Apr 17, 1990
- 4924823 Six-stroke internal combustion engine May 15, 1990
- 6253745 Multiple stroke engine having fuel and vapour charges Jul 3, 2001
- 6311651 Computer-controlled six-stroke internal combustion engine and its method of operation Nov 6, 2001
- 6571749 Computer-controlled six-stroke cycle internal combustion engine and its method of operation Jun 3, 2003
- 7021272 Computer controlled multi-stroke cycle power generating assembly and method of operation Apr 4, 2006.

Diesel Engine

A diesel engine (also known as a compression-ignition engine) is an internal combustion engine that uses the heat of compression to initiate ignition to burn the fuel, which is injected into the combustion chamber. This is in contrast to spark-ignition engines such as a petrol engine (gasoline engine) or gas engine (using a gaseous fuel as opposed to gasoline), which uses a spark plug to ignite an air-fuel mixture. The engine was developed by Rudolf Diesel in 1893.

The diesel engine has the highest thermal efficiency of any regular internal or external combustion engine due to its very high compression ratio. Low-speed Diesel engines (as used in ships and other applications where overall engine weight is relatively unimportant) often have a thermal efficiency which exceeds 50 percent.

Diesel engines are manufactured in two-stroke and four-stroke versions. They were originally used as a more efficient replacement for stationary steam engines. Since the 1910s they have been used in submarines and ships. Use in locomotives, trucks, heavy equipment and electric generating plants followed later. In the 1930s, they slowly began to be used in a few automobiles. Since the 1970s, the use of diesel engines in larger on-road and off-road vehicles in the USA increased. As of 2007, about 50 percent of all new car sales in Europe are diesel.

The world's largest diesel engine is currently a Wärtsilä Sulzer RT96-C Common Rail marine diesel of about 108,920 hp (81,220 kW) @ 102 rpm

History

Rudolf Diesel was born in Paris in 1858 into a family of German expatriates. He was educated at Munich Polytechnic. After graduation he was employed as a refrigerator engineer, but his true love lay in engine design. Diesel designed many heat engines, including a solar-powered air engine.

In 1892 he received patents in Germany, Switzerland, the United Kingdom and filed in the United States for "Method of and Apparatus for Converting Heat into Work".

In 1893 he described a "slow-combustion engine" that first compressed air thereby raising its temperature above the igniting-point of the fuel, then gradually introducing fuel while letting the mixture expand "against resistance sufficiently to prevent an essential increase of temperature and pressure", then cutting off fuel and "expanding without transfer of heat".

In 1894 and 1895 he filed patents and addenda in various countries for his Diesel engine; the first patents were issued in Spain (No.16,654), France (No.243,531) and Belgium (No.113,139) in December 1894, and in Germany (No.86,633) in 1895 and the United States (No.608,845) in 1898. He operated his first successful engine in 1897. His engine was the first to prove that fuel could be ignited solely with high compression.

Though best known for his invention of the pressure-ignited heat engine that bears his name, Rudolf Diesel was also a well-respected thermal engineer and a social theorist. Diesel's inventions have three points in common: they relate to heat transfer by natural physical processes or laws; they involve markedly creative mechanical design; and they were initially motivated by the inventor's concept of sociological needs.

Rudolf Diesel originally conceived the diesel engine to enable independent craftsmen and artisans to compete with industry.

At Augsburg, on August 10, 1893, Rudolf Diesel's prime model, a single 10-foot (3.0 m) iron cylinder with a flywheel at its base, ran on its own power for the first time. Diesel spent two more years making improvements and in 1896 demonstrated another model with a theoretical efficiency of 75 percent, in contrast to the 10 percent efficiency of the steam engine.

By 1898, Diesel had become a millionaire. His engines were used to power pipelines, electric and water plants, automobiles and trucks, and marine craft. They were soon to be used in mines, oil fields, factories, and transoceanic shipping.

How Diesel Engines Work

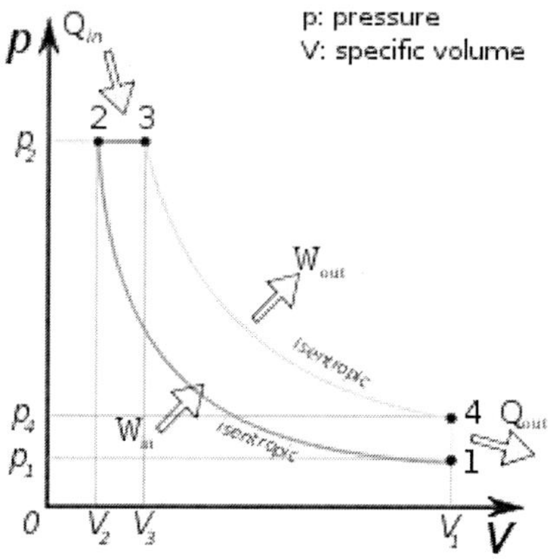

Figure 2: p-V Diagram for the Ideal Diesel cycle. The cycle follows the numbers 1-4 in clockwise direction. In the diesel cycle the combustion occurs at almost constant pressure and the exhaust occurs at constant volume. On this diagram the work that is generated for each cycle corresponds to the area within the loop.

Figure 3: Diesel engine model, left side

Figure 4: Diesel engine model, right side

The diesel internal combustion engine differs from the gasoline powered Otto cycle by using highly compressed hot air to ignite the fuel rather than using a spark plug (*compression ignition* rather than *spark ignition*).

In the true diesel engine, only air is initially introduced into the combustion chamber. The air is then compressed with a compression ratio typically between 15:1 and 22:1 resulting in 40-bar (4.0 MPa; 580 psi) pressure compared to 8 to 14 bars (0.80 to 1.4 MPa) (about 200 psi) in the petrol engine. This high compression heats the air to 550 °C (1,022 °F). At about the top of the compression stroke, fuel is injected directly into the compressed air in the combustion chamber. This may be into a (typically toroidal) void in the top of the piston or a *pre-chamber* depending upon the design of the engine. The fuel injector ensures that the fuel is broken down into small droplets, and that the fuel is distributed evenly.

The heat of the compressed air vapourises fuel from the surface of the droplets. The vapour is then ignited by the heat from the compressed air in the combustion chamber, the droplets continue to vapourise from their surfaces and burn, getting smaller, until all the fuel in the droplets has been burnt. The start of vapourisation causes a delay period during ignition and the characteristic diesel knocking sound as the vapour reaches ignition temperature and causes an abrupt increase in pressure above the piston. The rapid expansion of combustion gases then drives the piston downward, supplying power to the crankshaft. Engines for scale-model aeroplanes use a variant of the Diesel principle but premix fuel and air via a carburation system external to the combustion chambers.

As well as the high level of compression allowing combustion to take place without a separate ignition system, a high compression ratio greatly increases the engine's efficiency. Increasing the compression ratio in a spark-ignition engine where fuel and air are mixed before entry to the cylinder is limited by the need to prevent damaging pre-ignition. Since only air is compressed in a diesel engine, and fuel is not introduced into the cylinder until shortly before top dead centre (TDC), premature detonation is not an issue and compression ratios are much higher.

Early Fuel Injection Systems

Diesel's original engine injected fuel with the assistance of compressed air, which atomized the fuel and forced it into the engine through a nozzle (a similar principle to an aerosol spray). The nozzle opening was closed by a pin valve lifted by the camshaft to initiate the fuel injection before top dead centre (TDC). This is called an airblast injection. Driving the three stage compressor used some power but the efficiency and net power output was more than any other combustion engine at that time.

Diesel engines in service today raise the fuel to extreme pressures by mechanical pumps and deliver it to the combustion chamber by pressure-activated injectors without compressed air. With direct injected diesels, injectors spray fuel through 4 to 12 small orifices in its nozzle. The early air injection diesels always had a superior combustion without the sharp increase in pressure during combustion. Research is now being performed and patents are being taken out to again use some form of air injection to reduce the nitrogen oxides and pollution, reverting to Diesel's original implementation with its superior combustion and possibly quieter operation.

In all major aspects, the modern diesel engine holds true to Rudolf Diesel's original design, that of igniting fuel by compression at an extremely high pressure within the cylinder. With much higher pressures and high technology injectors, present-day diesel engines use the so-called solid injection system applied by Herbert Akroyd Stuart for his hot bulb engine. The indirect injection engine could be considered the latest development of these low speed *hot bulb* ignition engines..

Fuel Delivery

A vital component of all diesel engines is a mechanical or electronic governor which regulates the idling speed and maximum speed of the engine by controlling the rate of fuel delivery. Unlike Otto-cycle engines, incoming air is not throttled and a diesel engine without a governor cannot have a stable idling speed and can easily overspeed, resulting in its destruction. Mechanically governed fuel injection systems are driven by the engine's gear train. These systems use a combination of springs and weights to control fuel delivery relative to both load and speed. Modern electronically controlled diesel engines control fuel delivery by use of an electronic control module (ECM) or electronic control unit (ECU).

The ECM/ECU receives an engine speed signal, as well as other operating parameters such as intake manifold pressure and fuel temperature, from a sensor and controls the amount of fuel and start of injection timing through actuators to maximise power and efficiency and minimise emissions. Controlling the timing of the start of injection of fuel into the cylinder is a key to minimising emissions, and maximising fuel economy (efficiency), of the engine. The timing is measured in degrees of crank angle of the piston before top dead centre. For example, if the ECM/ECU initiates fuel injection when the piston is 10 degrees before TDC, the start of injection, or timing, is said to be 10° BTDC. Optimal timing will depend on the engine design as well as its speed and load.

Advancing the start of injection (injecting before the piston reaches to its SOI-TDC) results in higher in-cylinder pressure and temperature, and higher efficiency, but also results in elevated engine noise and increased oxides of nitrogen (NO_x) emissions due to higher combustion temperatures. Delaying start of injection causes incomplete combustion, reduced fuel efficiency and an increase in exhaust smoke, containing a considerable amount of particulate matter and unburned hydrocarbons.

Major Advantages

Diesel engines have several advantages over other internal combustion engines:

- They burn less fuel than a petrol engine performing the same work, due to the engine's higher temperature of combustion and greater expansion ratio. Gasoline engines are typically 30 percent efficient while diesel engines can convert over 45 percent of the fuel energy into mechanical energy.
- They have no high voltage electrical ignition system, resulting in high reliability and easy adaptation to damp environments. The absence of coils, spark plug wires, etc., also eliminates a source of radio frequency emissions which can interfere with navigation and communication equipment, which is especially important in marine and aircraft applications.
- The life of a diesel engine is generally about twice as long as that of a petrol engine due to the increased strength of parts used. Diesel fuel has better lubrication properties than petrol as well.
- Diesel fuel is distilled directly from petroleum. Distillation yields some gasoline, but the yield would be inadequate without catalytic reforming, which is a more costly process.
- Diesel fuel is considered safer than petrol in many applications. Although diesel fuel will burn in open air using a wick, it will not explode and does not release a large amount of flammable vapour. The low vapour pressure of diesel is especially

advantageous in marine applications, where the accumulation of explosive fuel-air mixtures is a particular hazard. For the same reason, diesel engines are immune to vapour lock.

- For any given partial load the fuel efficiency (mass burned per energy produced) of a diesel engine remains nearly constant, as opposed to petrol and turbine engines which use proportionally more fuel with partial power outputs.
- They generate less waste heat in cooling and exhaust.
- Diesel engines can accept super- or turbo-charging pressure without any natural limit, constrained only by the strength of engine components. This is unlike petrol engines, which inevitably suffer detonation at higher pressure.
- The carbon monoxide content of the exhaust is minimal, therefore diesel engines are used in underground mines.
- Biodiesel is an easily synthesized, non-petroleum-based fuel (through transesterification) which can run directly in many diesel engines, while gasoline engines either need adaptation to run synthetic fuels or else use them as an additive to gasoline (e.g., ethanol added to gasohol).

Mechanical and Electronic Injection

Many configurations of fuel injection have been used over the past century (1901–2000).

Most present day (2008) diesel engines make use of a camshaft, rotating at half crankshaft speed, lifted mechanical single plunger high-pressure fuel pump driven by the engine crankshaft. For each engine cylinder, the corresponding plunger in the fuel pump measures out the correct amount of fuel and determines the timing of each injection. These engines use injectors that are very precise spring-loaded valves that open and close at a specific fuel pressure. Separate high-pressure fuel lines connect the fuel pump with each cylinder. Fuel volume for each single combustion is controlled by a slanted groove in the plunger which rotates only a few degrees releasing the pressure and is controlled by a mechanical governor, consisting of weights rotating at engine speed constrained by springs and a lever. The injectors are held open by the fuel pressure. On high-speed engines the plunger pumps are together in one unit. The length of fuel lines from the pump to each injector is normally the same for each cylinder in order to obtain the same pressure delay.

A cheaper configuration on high-speed engines with fewer than six cylinders is to use an axial-piston distributor pump, consisting of one rotating pump plunger delivering fuel to a valve and line for each cylinder (functionally analogous to points and distributor cap on an Otto engine).

Many modern systems have a single fuel pump which supplies fuel constantly at high pressure with a common rail (single fuel line common) to each injector. Each injector has a solenoid operated by an electronic control unit, resulting in more accurate control of injector opening times that depend on other control conditions, such as engine speed and loading, and providing better engine performance and fuel economy. This design is also mechanically simpler than the combined pump and valve design, making it generally more reliable, and less loud, than its mechanical counterpart. This system does have have the drawback of requiring a reliable electrical system for operation.

Both mechanical and electronic injection systems can be used in either direct or indirect injection configurations.

Older diesel engines with mechanical injection pumps could be inadvertently run in reverse, albeit very inefficiently. When this occurs, massive amounts of soot are ejected from the air intake. This was often a consequence of push starting a vehicle using the wrong gear. Large ship diesels are capable of running either direction.

Indirect Injection

In an internal combustion engine, the term indirect injection refers to a fuel injection where fuel is not directly injected into the combustion chamber. Gasoline engines are usually equipped with indirect injection systems, wherein a fuel injector delivers the fuel at some point before the intake valve.

An indirect injection diesel engine delivers fuel into a chamber off the combustion chamber, called a prechamber, where combustion begins and then spreads into the main combustion chamber. The prechamber is carefully designed to ensure adequate mixing of the atomized fuel with the compression-heated air. This has the effect of slowing the rate of combustion, which tends to reduce audible noise and softens the shock of combustion and produces lower stresses on the engine components. The addition of a prechamber, however, increases heat loss to the cooling system and thereby lowers engine efficiency. The engine requires glow plugs for starting. In an indirect injection system the air moves fast, mixing the fuel and air. This simplifies injector design and allows the use of smaller engines and less tightly toleranced designs which are simpler to manufacture and more reliable. Direct injection, by contrast, uses slow-moving air and fast-moving fuel; both the design and manufacture of the injectors is more difficult. The optimisation of the in-cylinder air flow is much more difficult than designing a prechamber. There is much more integration between the design of the injector and the engine. It is for this reason that car diesel engines were almost all indirect injection until the ready availability of powerful CFD simulation systems made the adoption of direct injection practical.

Classification of Indirect Combustion Chambers (Prechambers)

Swirl Chamber

It consists of a spherical chamber located in the cylinder head and separated from the engine cylinder by a tangential throat. About 50% of the air enters the swirl chamber during the compression stroke of the engine, producing a swirl.

After combustion, the products return through the same throat to the main cylinder at much higher velocity. So more heat loss to walls of the passage takes place. This type of chamber finds application in engines in which fuel control and engine stability are more important than fuel economy. These are Ricardo chambers.

Precombustion Chamber

This chamber is located at the cylinder head and is connected to the engine cylinder by small holes. It occupies 40% of the total cylinder volume. During the compression stroke, air from the main cylinder enters the precombustion chamber. At this moment, fuel is injected into the precombustion chamber and combustion begins. Pressure increases and the fuel droplets are forced through the small holes into the main cylinder, resulting in a very good mix of the fuel and air. The bulk of the combustion actually takes place in the main cylinder. This type of combustion chamber has multi-fuel capability because the temperature of the prechamber vapourises the fuel before the main combustion event occurs.

Air Cell Chamber

The air cell is a small cylindrical chamber with a hole in one end. It is mounted more or less coaxially with the injector, said axis being parallel to the piston crown, with the injector firing across a small cavity which is open to the cylinder into the hole in the end of the air cell. The air cell is mounted so as to minimise thermal contact with the mass of the head. A pintle injector with a narrow spray pattern is used. At TDC the majority of the charge mass is contained in the cavity and air cell.

When the injector fires the jet of fuel enters the air cell and ignites. This results in a jet of flame shooting back out of the air cell directly into the jet of fuel still issuing from the injector. The heat and turbulence give excellent fuel vapourisation and mixing properties. Also since the majority of the combustion takes place outside the air cell in the cavity, which communicates directly with the cylinder, there is less heat loss involved in transferring the burning charge into the cylinder.

Air cell injection can be considered as a sort of half way stage between fully indirect and fully direct injection, gaining some of the efficiency advantages of direct injection while retaining the simplicity and ease of development of indirect injection.

Air cell chambers are commonly named Lanova air chambers.

Advantages of Indirect Injection Combustion Chambers

1. Smaller diesels can be produced.
2. The injection pressure required is low, so the injector is cheaper to produce.
3. The injection direction is of less importance.
4. Indirect injection is much simpler to design and manufacture; less injector development is required and the injection pressures are low (1500 psi versus 5000 psi and higher for direct injection)
5. The lower stresses that indirect injection imposes on internal components mean that it is possible to produce petrol and indirect injection diesel versions of the same basic engine. At best such types differ only in the cylinder head and the need to fit a distributor and spark plugs in the petrol version whilst fitting an injection pump and injectors to the diesel. Examples include the BMC A-Series and B-Series engines and the Land Rover 2.25/2.5-litre 4-cylinder types. Such designs allow petrol and diesel versions of the same vehicle to be built with minimal design changes between them.
6. Higher engine speeds can be reached, since burning continues in the prechamber. The Mercedes-Benz type prechamber is able to achieve over 6000rpm in a turbocharged engine.

7. Indirect injection is superior for running on thicker vegetable oil fuel, due to lower pressure injection, a longer burn time and increased swirl ensuring more complete combustion.

Disadvantages

1. Specific fuel consumption is high because of heat loss due to large exposed areas and pressure loss due to air motion through the throats.
2. Glowplugs are needed for a cold engine start.
3. Because the heat and pressure of combustion is applied to one specific point on the piston as it exits the precombustion chamber or swirl chamber, such engines are less suited to high specific power outputs (such as turbocharging or tuning) than direct injection diesels. The increased temperature and pressure on one part of the piston crown causes uneven expansion which can lead to cracking, distortion or other damage.

Maintenance Hazards

Fuel injection introduces potential hazards in engine maintenance due to the high fuel pressures used. Residual pressure can remain in the fuel lines long after an injection-equipped engine has been shut down. This residual pressure must be relieved, and if it is done so by external bleed-off, the fuel must be safely contained. If a high-pressure diesel fuel injector is removed from its seat and operated in open air, there is a risk to the operator of injury by hypodermic jet-injection, even with only 100 psi pressure. The first known such injury occurred in 1937 during a diesel engine maintenance operation.

Direct Injection

Direct injection diesel engines have injectors mounted at the top of the combustion chamber. The injectors are activated using one of two methods - hydraulic pressure from the fuel pump, or an electonic signal from an engine controller.

Hydraulic pressure activated injectors can produce harsh engine noise. Fuel consumption was about 15 to 20 percent lower than indirect injection diesels. The extra noise was generally not a problem for industrial uses of the engine. But for automotive usage, buyers had to decide whether or not the increased fuel efficiency would compensate for the extra noise.

Electronic control of the fuel injection transformed the direct injection engine. This was pioneered by Fiat in 1986 (Croma). The injection pressure remained around 300 bar (30 MPa; 4,400 psi), but the injection timing fuel quantity, EGR, and turbo boost are all electronically controlled. This gives more precise control of these parameters, resulting in lowered emissions and quieter, smoother running engines.

Unit Injector

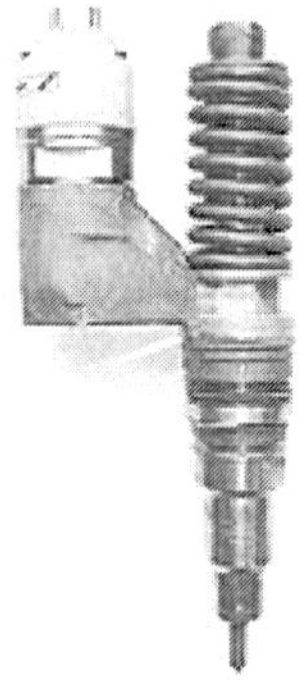

Figure 5: Early Lucas electronic diesel unit injector

Unit Injector (UI) is an integrated direct fuel injection system for diesel engines, combining the injector nozzle and the injection pump in a single component. The pump is usually driven by a shared camshaft.

History

In 1911 a patent was issued in Great Britain for a unit injector resembling those in use today to Frederick Lamplough.

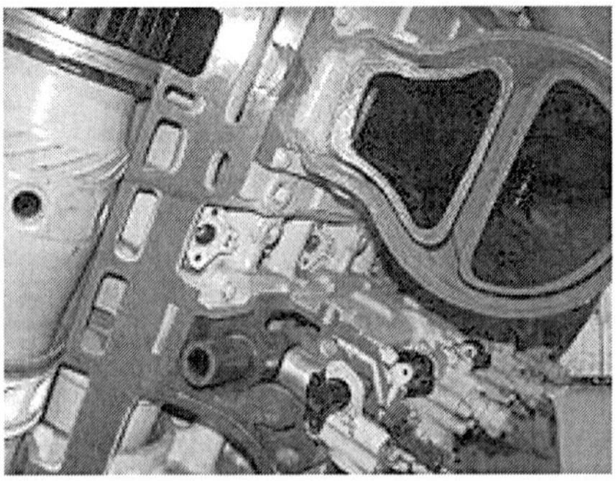

Figure 6: Napier Deltic opposed-piston two-stroke, sectioned. The unit injectors are low down, below the yellow fuel passages, driven by a camshaft to their left and injecting into the centre of the cylinder liner (pale blue).

Commercial usage of unit injectors in the U.S. began in early 1930s on Winton engines powering locomotives, boats, even US Navy submarines,, and in 1934, Arthur Fielden was granted U.S. patent No.1,981,913 on the unit injector design later used for the General Motors two-stroke diesel engines. Most mid-sized diesel engines used a single pump and separate injectors, but some makers, such as Detroit Diesel became well-known for favouring unit injectors.

In 1994, Robert Bosch GmbH supplied the first electronic Unit Injector for commercial vehicles, and other manufacturers soon followed. Today, major manufacturers include Robert Bosch GmbH, CAT, Cummins, Delphi Corp., Detroit Diesel and the Delphi Corp. acquired Lucas Automotive.

Design and Technology

Design of the Unit Injector eliminates the need for high pressure fuel pipes, and with that their associated failures, as well as allowing for much higher injection pressure to occur. The unit injector system allows accurate injection timing, and amount control as in the common rail system.

The Unit Injector is fitted into the engine cylinder head, where the fuel is supplied via integral ducts machined directly into the cylinder head. Each injector has its own pumping element, and in the case of electronic control, a fuel solenoid valve as well. The fuel system is divided into the low pressure (<500 kPa) fuel supply system, and the high pressure injection system (<2000 bar).

Operation Principle

Figure 7: Delphi E1 UI on the Volvo D13A engine

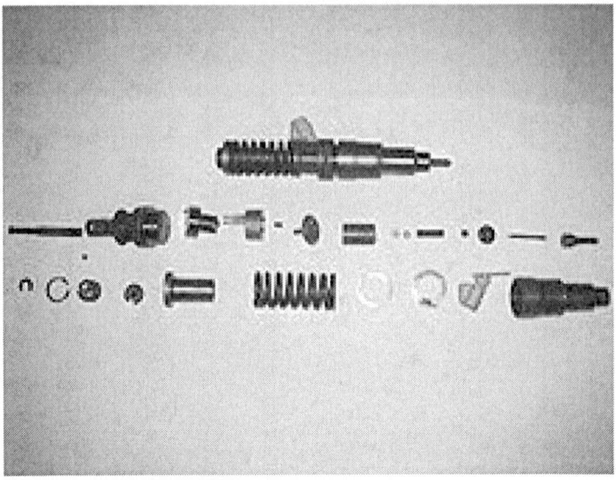

Figure 8: Delphi E1 UI parts

The basic operation can be described as a sequence of four separate phases: the *filling phase*, the *spill phase*, the *injection phase*, and the *pressure reduction phase.*

A low pressure fuel delivery pump supplies filtered diesel fuel into the cylinder head fuel ducts, and into each injector fuel port of constant stroke pump plunger injector, which is overhead camshaft operated.

Fill Phase : The constant stroke pump element on the way up draws fuel from the supply duct in to the chamber, and as long as electric solenoid valve remains de-energised fuel line is open.

Spill Phase: The pump element is on the way down, and as long as solenoid valve remains de-energised the fuel line is open and fuel flows in through into the return duct.

Injection Phase: The pump element is still on the way down, the solenoid is now energised and fuel line is now closed. The fuel can not pass back into return duct, and is now compressed by the plunger until pressure exceeds specific "opening" pressure, and the injector nozzle needle lifts, allowing fuel to be injected into the combustion chamber.

Pressure Reduction Phase : The plunger is still on its way down, the engine ECU de-energises the solenoid when required quantity of fuel is delivered, the fuel valve opens, fuel can flow back into return duct, causing pressure drop, which in turn causes the injector nozzle needle to shut, hence no more fuel is injected.

The start of an injection is controlled by the solenoid closing point, and the injected fuel quantity is determined by the closing time, which is the length of time the solenoid remains closed. The solenoid operation is fully controlled by the engine ECU.

Additional Functions

The use of electronic control allows for special functions; such as temperature controlled injection timing, cylinder balancing (smooth idle), switching off individual cylinders under part load for further reduction in emissions and fuel consumption, and multi-pulse injection (more than one injection occurrence during one engine cycle).

Further Development and Applications

In 1993, CAT introduced "Hydraulically-actuated Electronic Unit Injection" (HEUI), where the injectors are no longer camshaft operated. First available on Navistar's 7.3 litres (445.5 cu in), V8 diesel engine. HEUI uses engine oil pressure to power high pressure fuel injection, where usual method of unit injector operation is the engine camshaft.

Unit injector fuel systems are being used on wide variety of vehicles and engines; commercial vehicle from manufacturers such as Volvo, Cummins, Detroit Diesel, CAT, and passenger vehicles from manufacturers such as Land Rover and Volkswagen Group, among others.

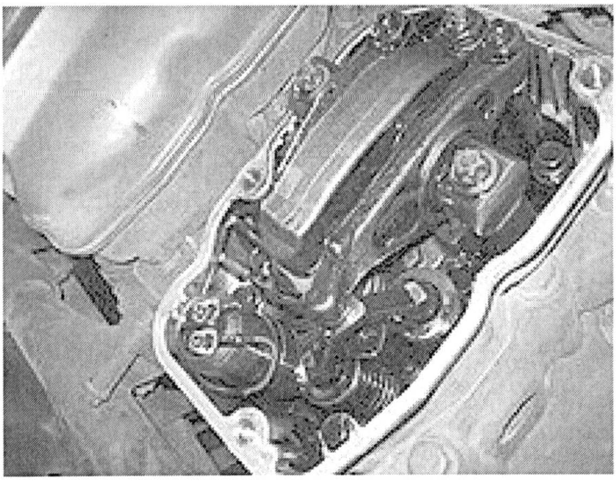

Figure 9: Bosch UI on Scania R164 V8

The Volkswagen Group mainstream marques use unit injector systems in their Suction Diesel Injection (SDI) and Turbocharged Direct Injection (TDI) diesel engines.

- In North America, the Volkswagen Jetta, Golf, and New Beetle TDI 2004-2006 are Mk4 Pumpe Düse (newer models use Mk5 BEW engines and older models use ALH engines).

Volkswagen Group major-interest truck and diesel engine maker Scania AB also use the unit injector system, which they call "Pumpe-Düse-Einspritzung", or "PDE".

Common Rail

Common rail direct fuel injection is a modern variant of direct fuel injection system for petrol and diesel engines.

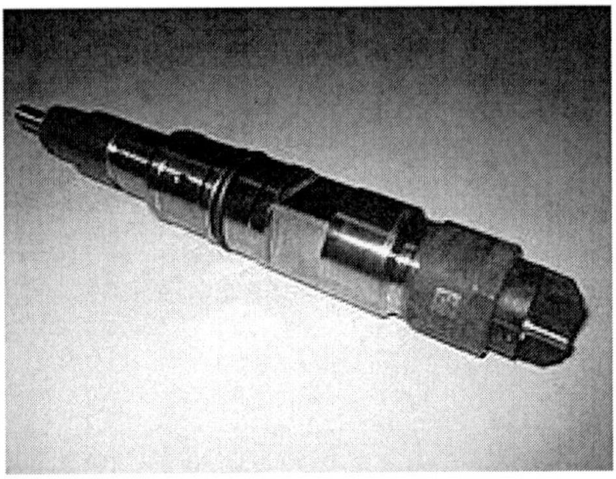

Figure 10: Common rail fuel injector

On diesel engines, it features a high-pressure (over 1,000 bar/ 15,000 psi) fuel rail feeding individual solenoid valves, as opposed to low-pressure fuel pump feeding unit injectors (pump nozzles). Third-generation common rail diesels now feature piezoelectric injectors for increased precision, with fuel pressures up to 1,800 bar/26,000 psi.

In gasoline engines, it is used in gasoline direct injection engine technology.

History

The common rail system prototype was developed in the late 1960s by Robert Huber of Switzerland and the technology further developed by Dr. Marco Ganser at the Swiss Federal Institute of Technology in Zurich, later of Ganser-Hydromag AG (est.1995) in Oberägeri.

The first successful usage in production vehicle began in Japan by the mid-1990s. Dr. Shohei Itoh and Masahiko Miyaki of the Denso Corporation, a Japanese automotive parts manufacturer, developed the common rail fuel system for heavy duty vehicles and turned it into practical use on their ECD-U2 common-rail system mounted on the Hino Rising Ranger truck and sold for general use in 1995. Denso claims the first commercial high pressure common rail system in 1995.

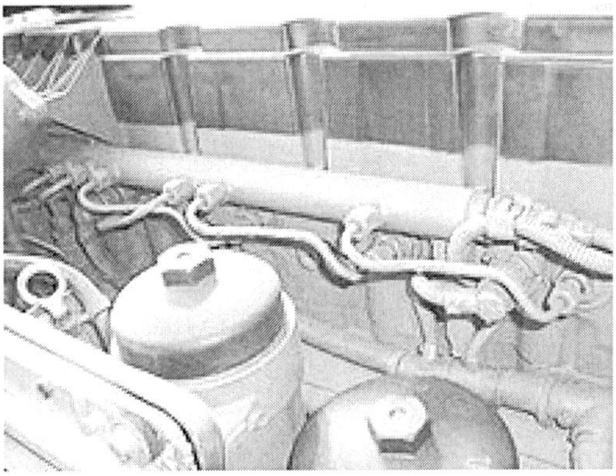

Figure 11: Common rail fuel system close up

Modern common rail systems, whilst working on the same principle, are governed by an engine control unit (ECU) which opens each injector electronically rather than mechanically. This was extensively prototyped in the 1990s with collaboration between Magneti Marelli, Centro Ricerche Fiat and Elasis. After research and development by the Fiat Group, the design was acquired by the German company Robert Bosch GmbH for completion of development and refinement for mass-production. In hindsight the sale appeared to be a tactical error for Fiat as the new technology proved to be highly profitable. The company had little choice but to sell, however, as it was in a poor financial state at the time and lacked the resources to complete development on its own. In 1997 they extended its use for passenger cars. The first passenger car that used the common rail system was the 1997 model Alfa Romeo 156 2.4 JTD, and later on that same year Mercedes-Benz C 220 CDI.

Common rail engines have been used in marine and locomotive applications for some time. The Cooper-Bessemer GN-8 (circa 1942) is an example of a hydraulically operated common rail diesel engine, also known as a modified common rail.

Vickers used common rail systems in submarine engines circa 1916. Doxford Engines Ltd. (opposed piston heavy marine engines) used a common rail system (from 1921 to 1980) whereby a multi-cylinder reciprocating fuel pump generated a pressure of approximately 600bar with the fuel being stored in accumulator bottles. Pressure control was achieved by means of an adjustable pump discharge stroke and a "spill valve". Camshaft operated mechanical timing valves were used to supply the spring-loaded Brice/CAV/Lucas injectors which injected through the side of the cylinder into the chamber formed between the pistons. Early engines had a pair of timing cams, one for ahead running and one for astern. Later engines had two injectors per cylinder and the final series of constant pressure turbocharged engines were fitted with four injectors per cylinder. This system was used for the injection of both diesel oil and heavy fuel oil (600cSt heated to a temperature of approximately 130°C). The common rail system is suitable for all types of road cars with diesel engines, ranging from city cars such as the Fiat Nuova Panda to executive cars such as the Audi A6.

Common Rail Today

Today the common rail system has brought about a revolution in diesel engine technology. Robert Bosch GmbH, Delphi Automotive Systems, Denso Corporation, and Siemens VDO (now owned by Continental AG) are the main suppliers of modern common rail systems. The car makers refer to their common rail engines by their own brand names:

- BMW's D-engines (also used in the Land Rover Freelander TD4)
- Cummins and Scania's XPI (Developed under joint venture)
- Cummins CCR (Cummins pump with Bosch Injectors)
- Daimler's CDI (and on Chrysler's Jeep vehicles simply as CRD)
- Fiat Group's (Fiat, Alfa Romeo and Lancia) JTD (also branded as *MultiJet, JTDm, Ecotec CDTi, TiD, TTiD, DDiS, Quadra-Jet*)
- Ford Motor Company's TDCi Duratorq and Powerstroke
- General Motors Opel CDTI and earlier DTI
- General Motors Chevrolet VCDi (licensed from VM Motori; also branded as *Ecotec CDTI*)
- Honda's i-CTDi
- Hyundai-Kia's CRDi
- IKCO's EFD which is one of the members of the EF family. Supplier TBD
- Isuzu's iTEQ
- Komatsu's Tier3, Tier4, 4D95 *and higher - HPCR series Diesel engines.*
- Mahindra's CRDe
- Mazda's MZR-CD (1.4 MZ-CD, 1.6 MZ-CD manufactured by joint venture Ford/PSA Peugeot Citroën) and earlier DiTD

- Mitsubishi's DI-D (recently developed 4N1 engine family uses next generation 200 MPa (2000 bar) injection system))
- Nissan's dCi, Infiniti uses dCi engines, but not branded as dCi.
- Proton's SCDi
- PSA Peugeot Citroën's HDI or HDi (1.4HDI, 1.6 HDI, 2.0 HDI, 2.2 HDI and V6 HDI developed under joint venture with Ford)
- Renault's *dCi* and earlier dTi
- SsangYong's XDi (most of these engines are manufactured by Daimler AG)
- Subaru's Legacy TD (as of Jan 2008)
- Tata's DICOR & CR4
- Toyota's D-4D
- Volkswagen Group: The 6.0 V12 TDI, 4.2 TDI (V8), 2.7 and 3.0 TDI (V6), 1.6, 2.0 TDI (L4) and 1.2 TDI (L3) engines featured on current Seat, Skoda, VW and Audi models use common rail, as opposed to the earlier unit injector engines.
- Volvo 2.4D and D5 engines (1.6D, 2.0D manufactured by Ford and PSA Peugeot Citroen), Volvo Penta D-serie engines
- Wärtsilä-Sulzer 14RT-flex96-C "largest reciprocating engine in the world" designed by the Finnish manufacturer Wärtsilä

Principles

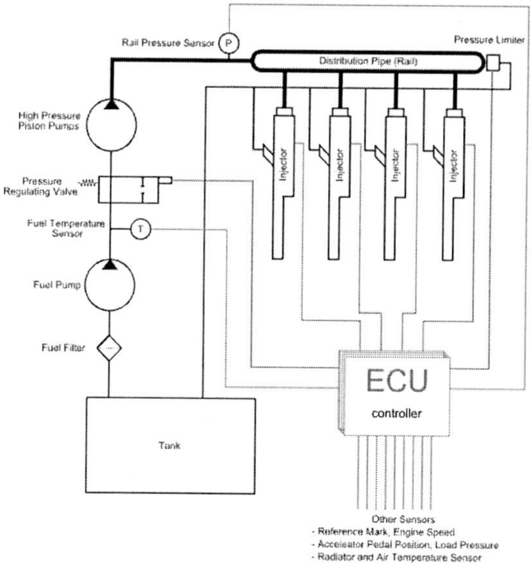

FIgure 12.

Solenoid or piezoelectric valves make possible fine electronic control over the fuel injection time and quantity and the higher pressure that the common rail technology makes available provides better fuel atomisation. In order to lower engine noise, the engine's electronic control unit can inject a small amount of diesel just before the main injection event ("pilot" injection), thus reducing its explosiveness and vibration, as well as optimising injection timing and quantity for variations in fuel quality, cold starting and so on. Some advanced common rail fuel systems perform as many as five injections per stroke.

Common rail engines require very short (< 1 s) or no heating up time at all and produce lower engine noise and emissions than older systems. Diesel engines have historically used various forms of fuel injection. Two common types include the unit injection system and the distributor/inline pump systems. While these older systems provided accurate fuel quantity and injection timing control, they were limited by several factors:

- They were cam driven and injection pressure was proportional to engine speed. This typically meant that the highest injection pressure could only be achieved at the highest engine speed and the maximum achievable injection pressure decreased as engine speed decreased. This relationship is true with all pumps, even those used on common rail systems; with the unit or distributor systems, however, the injection pressure is tied to the instantaneous pressure of a single pumping event with no accumulator and thus the relationship is more prominent and troublesome.
- They were limited in the number and timing of injection events that could be commanded during a single combustion event. While multiple injection events are possible with these older systems, it is much more difficult and costly to achieve.
- For the typical distributor/inline system, the start of injection occurred at a pre-determined pressure (often referred to as: pop pressure) and ended at a pre-determined pressure. This characteristic resulted from "dummy" injectors in the cylinder head which opened and closed at pressures determined by the spring preload applied to the plunger in the injector. Once the pressure in the injector reached a pre-determined level, the plunger would lift and injection would start.

In common rail systems, a high pressure pump stores a reservoir of fuel at high pressure — up to and above 2,000 bars (29,000 psi). The term "common rail" refers to the fact that all of the fuel injectors are supplied by a common fuel rail which is nothing more than a pressure accumulator where the fuel is stored at high pressure. This accumulator supplies multiple fuel injectors with high pressure fuel. This simplifies the purpose of the high pressure pump in that it only has to maintain a commanded pressure at a target (either mechanically or electronically controlled). The fuel injectors are typically ECU-controlled. When the fuel injectors are electrically activated, a hydraulic valve (consisting of a nozzle and plunger) is mechanically or hydraulically opened and fuel is sprayed into the cylinders at the desired pressure. Since the fuel pressure energy is stored remotely and the injectors are electrically actuated, the injection pressure at the start and end of injection is very near the pressure in the accumulator (rail), thus producing a square injection rate. If the accumulator, pump and plumbing are sized properly, the injection pressure and rate will be the same for each of the multiple injection events.

Cold Weather

Starting

In cold weather, high speed diesel engines can be difficult to start because the mass of the cylinder block and cylinder head absorb the heat of compression, preventing ignition due to the higher surface-to-volume ratio. Pre-chambered engines make use of small electric heaters inside the pre-chambers called glowplugs, while the direct-injected engines have these glowplugs in the combustion chamber. These engines also generally have a higher compression ratio of 19:1 to 21:1. Low-speed and compressed-air-started larger and intermediate-speed diesels do not have glowplugs and compression ratios are around 16:1.

Some engines (e.g., some Cummins models) use resistive grid heaters in the intake manifold to warm the inlet air until the engine reaches operating temperature. Engine block heaters (electric resistive heaters in the engine block) connected to the utility grid are often used when an engine is turned off for extended periods (more than an hour) in cold weather to reduce startup time and engine wear. Block heaters are also used for emergency power standby Diesel-powered generators which must rapidly pick up load on a power failure. In the past, a wider variety of cold-start methods were used. Some engines, such as Detroit Diesel engines and Lister-Petter engines, used a system to introduce small amounts of ether into the inlet manifold to start combustion.

Saab-Scania marine engines, Field Marshall tractors (among others) used slow-burning solid-fuel 'cigarettes' which were fitted into the cylinder head as a primitive glow plug.

Lucas developed the *Thermostart*, where an electrical heating element was combined with a small fuel valve in the inlet manifold. Diesel fuel slowly dripped from the valve onto the hot element and ignited. The flame heated the inlet manifold and when the engine was cranked, the flame was drawn into the cylinders to start combustion.

International Harvester developed a tractor in the 1930s that had a 7-litre 4-cylinder engine which started as a gasoline engine and ran on diesel after warming up.

The cylinder head had valves which opened for a portion of the compression stroke to reduce the effective compression ratio, and a magneto produced the spark. An automatic ratchet system automatically disengaged the ignition system and closed the valves once the engine had run for 30 seconds. The operator then switched off the petrol fuel system and opened the throttle on the diesel injection system.

Recent direct-injection systems are advanced to the extent that pre-chambers systems are not needed by using a common rail fuel system with electronic fuel injection.

Gelling

Diesel fuel is also prone to *waxing* or *gelling* in cold weather; both are terms for the solidification of diesel oil into a partially crystalline state.

The crystals build up in the fuel line (especially in fuel filters), eventually starving the engine of fuel and causing it to stop running. Low-output electric heaters in fuel tanks and around fuel lines are used to solve this problem.

Also, most engines have a *spill return* system, by which any excess fuel from the injector pump and injectors is returned to the fuel tank.

Once the engine has warmed, returning warm fuel prevents waxing in the tank. Due to improvements in fuel technology with additives, waxing rarely occurs in all but the coldest weather when a mix of diesel and kerosene should be used to run a vehicle.

Types—Size Groups

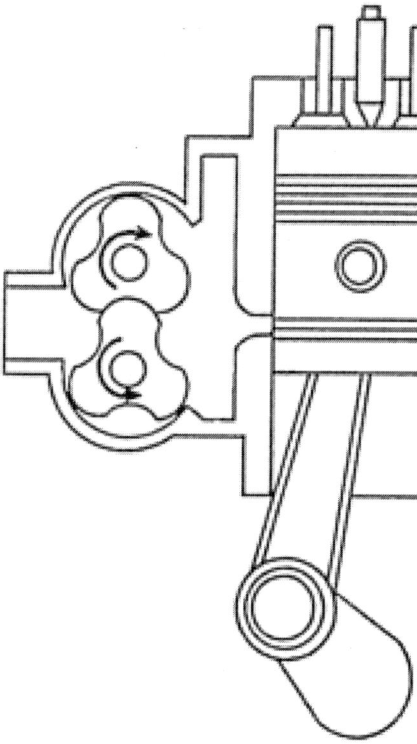

Figure 13: Two Cycle Diesel engine with Roots blower, typical of Locomotive Engines

There are three size groups of Diesel engines

- Small - Under 188 kW output
- Medium
- Large.

Basic Types of Diesel Engines

There are two basic types of Diesel Engines:

- Four Cycle
- Two Cycle.

Early

Rudolf Diesel based his engine on the design of the Gas engine created by Nikolaus Otto in 1876 with the goal of improving its efficiency. He patented his Diesel engine concepts in patents that were set forth in 1892 and 1893. As such, diesel engines in the late 19th and early 20th centuries used the same basic layout and form as industrial steam engines, with long-bore cylinders, external valve gear, cross-head bearings and an open crankshaft connected to a large flywheel.

Smaller engines would be built with vertical cylinders, while most medium- and large-sized industrial engines were built with horizontal cylinders, just as steam engines had been.

Engines could be built with more than one cylinder in both cases. The largest early diesels resembled the triple-expansion steam reciprocating engine, being tens of feet high with vertical cylinders arranged in-line.

These early engines ran at very slow speeds—partly due to the limitations of their air-blast injector equipment and partly so they would be compatible with the majority of industrial equipment designed for steam engines; maximum speeds of between 100 and 300 rpm were common. Engines were usually started by allowing compressed air into the cylinders to turn the engine, although smaller engines could be started by hand.

In 1897 when the first Diesel engine was completed Adolphus Busch travelled to Cologne and negotiated exclusive right to produce the Diesel engine in the USA and Canada. In his examination of the engine it was noted that the Diesel at that time operated at efficiencies of 32 to 35 percent thermodynamic efficiency when a typical triple expansion steam engine would operate at about 18 percent.

In the early decades of the 20th century, when large diesel engines were first being used, the engines took a form similar to the compound steam engines common at the time, with the piston being connected to the connecting rod by a crosshead bearing. Following steam engine practice some manufactures made double-acting two-stroke and four-stroke diesel engines to increase power output, with combustion taking place on both sides of the piston, with two sets of valve gear and fuel injection.

While it produced large amounts of power and was very efficient, the double-acting diesel engine's main problem was producing a good seal where the piston rod passed through the bottom of the lower combustion chamber to the crosshead bearing, and no more were built. By the 1930s turbochargers were fitted to some engines. Crosshead bearings are still used to reduce the wear on the cylinders in large long-stroke main marine engines.

Modern

Figure 14: A Yanmar 2GM20 marine diesel engine, installed in a sailboat

As with petrol engines, there are two classes of diesel engines in current use: two-stroke and four-stroke. The four-stroke type is the "classic" version, tracing its lineage back to Rudolf Diesel's prototype. It is also the most commonly used form, being the preferred power source for many motor vehicles, especially buses and trucks. Much larger engines, such as used for railroad locomotion and marine propulsion, are often two-stroke units, offering a more favourable power-to-weight ratio, as well as better fuel economy. The most powerful engines in the world are two-stroke diesels of mammoth dimensions.

Two-stroke diesel engine operation is similar to that of petrol counterparts, except that fuel is not mixed with air before induction, and the crankcase does not take an active role in the cycle. The traditional two-stroke design relies upon a mechanically driven positive displacement blower to charge the cylinders with air before compression and ignition. The charging process also assists in expelling (scavenging) combustion gases remaining from the previous power stroke. The archetype of the modern form of the two-stroke diesel is the Detroit Diesel engine, in which the blower pressurises a chamber in the engine block that is often referred to as the "air box". The (much larger) Electro-Motive prime mover used in EMD diesel-electric locomotives is built to the same principle.

In a two-stroke diesel engine, as the cylinder's piston approaches the bottom dead centre exhaust ports or valves are opened relieving most of the excess pressure after which a passage between the air box and the cylinder is opened, permitting air flow into the cylinder. The air flow blows the remaining combustion gases from the cylinder—this is the scavenging process. As the piston passes through bottom centre and starts upward, the passage is closed and compression commences, culminating in fuel injection and ignition. Refer to two-stroke diesel engines for more detailed coverage of aspiration types and supercharging of two-stroke diesel engines.

Normally, the number of cylinders are used in multiples of two, although any number of cylinders can be used as long as the load on the crankshaft is counterbalanced to prevent excessive vibration. The inline-six-cylinder design is the most prolific in light- to medium-duty engines, though small V8 and larger inline-four displacement engines are also common. Small-capacity engines (generally considered to be those below five litres in capacity) are generally four- or six-cylinder types, with the four-cylinder being the most common type found in automotive uses. Five-cylinder diesel engines have also been produced, being a compromise between the smooth running of the six-cylinder and the space-efficient dimensions of the four-cylinder. Diesel engines for smaller plant machinery, boats, tractors, generators and pumps may be four-, three- or two-cylinder types, with the single-cylinder diesel engine remaining for light stationary work. Direct reversible two-stroke marine diesels need at least three cylinders for reliable restarting forwards and reverse, while four-stroke diesels need at least six cylinders.

The desire to improve the diesel engine's power-to-weight ratio produced several novel cylinder arrangements to extract more power from a given capacity. The uniflow opposed-piston engine uses two pistons in one cylinder with the combustion cavity in the middle and gas in- and outlets at the ends. This makes a comparatively light, powerful, swiftly running and economic engine suitable for use in aviation. An example is the Junkers Jumo 204/205. The Napier Deltic engine, with three cylinders arranged in a triangular formation, each containing two opposed pistons, the whole engine having three crankshafts, is one of the better known.

Low-speed Diesels

Low-speed diesel engines (as used in ships and other applications where overall engine weight is relatively unimportant) often have a thermal efficiency which exceeds 50 percent.

Gas Generator

Before 1950, Sulzer started experimenting with two-stroke engines with boost pressures as high as 6 atmospheres, in which all the output power was taken from an exhaust gas turbine. The two-stroke pistons directly drove air compressor pistons to make a positive displacement gas generator. Opposed pistons were connected by linkages instead of crankshafts. Several of these units could be connected to provide power gas to one large output turbine. The overall thermal efficiency was roughly twice that of a simple gas turbine. This system was derived from Raúl Pateras Pescara's work on free-piston engines in the 1930s.

Advantages and Disadvantages versus Spark-ignition Engines

Power and Fuel Economy

The MAN S80ME-C7 low speed diesel engines use 155 gram fuel per kWh for an overall energy conversion efficiency of 54.4 percent, which is the highest conversion of fuel into power by any internal or external combustion engine. Diesel engines are more efficient than gasoline (petrol) engines of the same power rating, resulting in lower fuel consumption. A common margin is 40 percent more miles per gallon for an efficient turbodiesel. For example, the current model Škoda Octavia, using Volkswagen Group engines, has a combined Euro rating of 6.2 L/100 km (38 miles per US gallon, 16 km/L) for the 102 bhp (76 kW) petrol engine and 4.4 L/100 km (54 mpg, 23 km/L) for the 105 bhp (78 kW) diesel engine.

However, such a comparison does not take into account that diesel fuel is denser and contains about 15 percent more energy by volume. Although the calorific value of the fuel is slightly lower at 45.3 MJ/kg (megajoules per kilogram) than petrol at 45.8 MJ/kg, liquid diesel fuel is significantly denser than liquid petrol. This is significant because volume of fuel, in addition to mass, is an important consideration in mobile applications. No vehicle has an unlimited volume available for fuel storage.

Adjusting the numbers to account for the energy density of diesel fuel, the overall energy efficiency is still about 20 percent greater for the diesel version.

While a higher compression ratio is helpful in raising efficiency, diesel engines are much more efficient than gasoline (petrol) engines when at low power and at engine idle. Unlike the petrol engine, diesels lack a butterfly valve (throttle) in the inlet system, which closes at idle. This creates parasitic loss and destruction of availability of the incoming air, reducing the efficiency of petrol engines at idle. In many applications, such as marine, agriculture, and railways, diesels are left idling and unattended for many hours, sometimes even days. These advantages are especially attractive in locomotives.

The average diesel engine has a poorer power-to-weight ratio than the petrol engine. This is because the diesel must operate at lower engine speeds and because it needs heavier, stronger parts to resist the operating pressure caused by the high compression ratio of the engine and the large amounts of torque generated to the crankshaft. In addition, diesels are often built with stronger parts to give them longer lives and better reliability, important considerations in industrial applications.

For most industrial or nautical applications, reliability is considered more important than light weight and high power. Diesel fuel is injected just before the power stroke. As a result, the fuel cannot burn completely unless it has a sufficient amount of oxygen. This can result in incomplete combustion and black smoke in the exhaust if more fuel is injected than there is air available for the combustion process. Modern engines with electronic fuel delivery can adjust the timing and amount of fuel delivery (by changing the duration of the injection pulse), and so operate with less waste of fuel. In a mechanical system, the injection timing and duration must be set to be efficient at the anticipated operating rpm and load, and so the settings are less than ideal when the engine is running at any other RPM than what it is timed for. The electronic injection can "sense" engine revs, load, even boost and temperature, and continuously alter the timing to match the given situation. In the petrol engine, air and fuel are mixed for the entire compression stroke, ensuring complete mixing even at higher engine speeds.

Diesel engines usually have longer stroke lengths in order to achieve the necessary compression ratios. As a result piston and connecting rods are heavier and more force must be transmitted through the connecting rods and crankshaft to change the momentum of the piston. This is another reason that a diesel engine must be stronger for the same power output as a petrol engine.

Yet it is this characteristic that has allowed some enthusiasts to acquire significant power increases with turbocharged engines by making fairly simple and inexpensive modifications. A petrol engine of similar size cannot put out a comparable power increase without extensive alterations because the stock components cannot withstand the higher stresses placed upon them. Since a diesel engine is already built to withstand higher levels of stress, it makes an ideal candidate for performance tuning at little expense. However, it should be said that any modification that raises the amount of fuel and air put through a diesel engine will increase its operating temperature, which will reduce its life and increase service requirements. These are issues with newer, lighter, *high-performance* diesel engines which are not "overbuilt" to the degree of older engines and they are being pushed to provide greater power in smaller engines.

The addition of a turbocharger or supercharger to the engine greatly assists in increasing fuel economy and power output, mitigating the fuel-air intake speed limit mentioned above for a given engine displacement. Boost pressures can be higher on diesels than on petrol engines, due to the latter's susceptibility to knock, and the higher compression ratio allows a diesel engine to be more efficient than a comparable spark ignition engine. Because the burned gases are expanded further in a diesel engine cylinder, the exhaust gas is cooler, meaning turbochargers require less cooling, and can be more reliable, than with spark-ignition engines.

With a diesel, boost pressure is essentially unlimited. It is literally possible to run as much boost as the engine will physically stand before breaking apart.

The increased fuel economy of the diesel engine over the petrol engine means that the diesel produces less carbon dioxide (CO_2) per unit distance. Recent advances in production and changes in the political climate have increased the availability and awareness of biodiesel, an alternative to petroleum-derived diesel fuel with a much lower net-sum emission of CO_2, due to the absorption of CO_2 by plants used to produce the fuel. Although concerns are now being raised as to the negative effect this is having on the world food supply, as the growing of crops specifically for biofuels takes up land that could be used for food crops and uses water that could be used by both humans and animals. However, the use of waste vegetable oil, sawmill waste from managed forests in Finland, and advances in the production of vegetable oil from algae demonstrate great promise in providing feed stocks for sustainable biodiesel that are not in competition with food production.

Diesel engines have a lower rotational speed than an equivalent size petrol engine because the diesel-air mixture burns slower than the petrol-air mixture. A combination of improved mechanical technology (such as multi-stage injectors which fire a short "pilot charge" of fuel into the cylinder to warm the combustion chamber before delivering the main fuel charge), higher injection pressures that have improved the atomisation of fuel into smaller droplets, and electronic control (which can adjust the timing and length of the injection process to optimise it for all speeds and temperatures) have mitigated most of these problems in the latest generation of common-rail designs, while greatly improving engine efficiency. Poor power and narrow torque bands have been addressed by superchargers, turbochargers, (especially variable geometry turbochargers), intercoolers, and a large efficiency increase from about 35 percent for IDI to 45 percent for the latest engines in the last 15 years.

Even though diesel engines have a theoretical fuel efficiency of 75 percent , in practice it is lower. Engines in large diesel trucks, buses, and newer diesel cars can achieve peak efficiencies around 45 percent, and could reach 55 percent efficiency in the near future. However, average efficiency over a driving cycle is lower than peak efficiency. For example, it might be 37 percent for an engine with a peak efficiency of 44 percent.

Emissions

In diesel engines, conditions in the engine differ from the spark-ignition engine, since power is directly controlled by the fuel supply, rather than by controlling the air supply. Thus when the engine runs at low power, there is enough oxygen present to burn the fuel, and diesel engines only make significant amounts of carbon monoxide when running under a load.

Diesel exhaust is well known for its characteristic smell; but in Britain this smell in recent years has become much less because the sulfur is now removed from the fuel in the oil refinery.

Diesel exhaust has been found to contain a long list of toxic air contaminants. Among these pollutants, fine particle pollution is perhaps the most important as a cause of diesel's harmful health effects.

Power and Torque

For commercial uses requiring towing, load carrying and other tractive tasks, diesel engines tend to have better torque characteristics. Diesel engines tend to have their torque peak quite low in their speed range (usually between 1600 and 2000 rpm for a small-capacity unit, lower for a larger engine used in a truck). This provides smoother control over heavy loads when starting from rest, and, crucially, allows the diesel engine to be given higher loads at low speeds than a petrol engine, making them much more economical for these applications. This characteristic is not so desirable in private cars, so most modern diesels used in such vehicles use electronic control, variable geometry turbochargers and shorter piston strokes to achieve a wider spread of torque over the engine's speed range, typically peaking at around 2500–3000 rpm.

While diesel engines tend to have more torque at lower engine speeds than petrol engines, diesel engines tend to have a narrower power band than petrol engines. Naturally aspirated diesels tend to lack power and torque at the top of their speed range. This narrow band is a reason why a vehicle such as a truck may have a gearbox with as many as 18 or more gears, to allow the engine's power to be used effectively at all speeds.

Turbochargers tend to improve power at high engine speeds; superchargers improve power at lower speeds; and variable geometry turbochargers improve the engine's performance equally by flattening the torque curve.

Noise

The characteristic noise of a diesel engine is variably called diesel clatter, diesel nailing, or diesel knock. Diesel clatter is caused largely by the diesel combustion process; the sudden ignition of the diesel fuel when injected into the combustion chamber causes a pressure wave. Engine designers can reduce diesel clatter through: indirect injection; pilot or pre-injection; injection timing; injection rate; compression ratio; turbo boost; and exhaust gas recirculation (EGR).

Common rail diesel injection systems permit multiple injection events as an aid to noise reduction. Diesel fuels with a higher cetane rating modify the combustion process and reduce diesel clatter. CN (Cetane number) can be raised by distilling higher quality crude oil, by catalysing a higher quality product or by using a cetane improving additive. Some oil companies market high cetane or premium diesel. Biodiesel has a higher cetane number than petrodiesel, typically 55CN for 100% biodiesel.

A combination of improved mechanical technology such as multi-stage injectors which fire a short "pilot charge" of fuel into the cylinder to initiate combustion before delivering the main fuel charge, higher injection pressures that have improved the atomisation of fuel into smaller droplets, and electronic control (which can adjust the timing and length of the injection process to optimise it for all speeds and temperatures), have partially mitigated these problems in the latest generation of common-rail designs, while improving engine efficiency.

Reliability

The lack of an electrical ignition system greatly improves the reliability. The high durability of a diesel engine is also due to its overbuilt nature, a benefit that is magnified by the lower rotating speeds in diesels. Diesel fuel is a better lubricant than petrol so is less harmful to the oil film on piston rings and cylinder bores; it is routine for diesel engines to cover 250,000 miles (400,000 km) or more without a rebuild.

Due to the greater compression force required and the increased weight of the stronger components, starting a diesel engine is harder. More torque is required to push the engine through compression.

Either an electrical starter or an air-start system is used to start the engine turning. On large engines, pre-lubrication and slow turning of an engine, as well as heating, are required to minimise the amount of engine damage during initial start-up and running. Some smaller military diesels can be started with an explosive cartridge, called a Coffman starter, which provides the extra power required to get the machine turning.

In the past, Caterpillar and John Deere used a small petrol *pony* engine in their tractors to start the primary diesel engine. The pony engine heated the diesel to aid in ignition and used a small clutch and transmission to spin up the diesel engine. Even more unusual was an International Harvester design in which the diesel engine had its own carburetor and ignition system, and started on petrol. Once warmed up, the operator moved two levers to switch the engine to diesel operation, and work could begin. These engines had very complex cylinder heads, with their own petrol combustion chambers, and were vulnerable to expensive damage if special care was not taken (especially in letting the engine cool before turning it off).

Quality and Variety of Fuels

Petrol/gasoline engines are limited in the variety and quality of the fuels they can burn. Older petrol engines fitted with a carburetor required a volatile fuel that would vapourise easily to create the necessary air-fuel ratio for combustion. Because both air and fuel are admitted to the cylinder, if the compression ratio of the engine is too high or the fuel too volatile (with too low an octane rating), the fuel will ignite under compression, as in a diesel engine, before the piston reaches the top of its stroke. This pre-ignition causes a power loss and over time major damage to the piston and cylinder. The need for a fuel that is volatile enough to vapourise but not too volatile (to avoid pre-ignition) means that petrol engines will only run on a narrow range of fuels. There has been some success at dual-fuel engines that use petrol and ethanol, petrol and propane, and petrol and methane.

In diesel engines, a mechanical injector system vapourises the fuel directly into the combustion chamber or a pre-combustion chamber. This *forced vapourisation* means that less-volatile fuels can be used. More crucially, because only air is inducted into the cylinder in a diesel engine, the compression ratio can be much higher as there is no risk of pre-ignition provided the injection process is accurately timed. This means that cylinder temperatures are much higher in a diesel engine than a petrol engine, allowing less volatile fuels to be used.

Diesel fuel is a form of light fuel oil, very similar to kerosene/ paraffin, but diesel engines, especially older or simple designs that lack precision electronic injection systems, can run on a wide variety of other fuels. Some of the most common alternatives are Jet A-1 type jet fuel or vegetable oil from a very wide variety of plants. Some engines can be run on vegetable oil without modification, and most others require fairly basic alterations. Biodiesel is a pure diesel-like fuel refined from vegetable oil and can be used in nearly all diesel engines. Requirements for fuels to be used in diesel engines are the ability of the fuel to flow along the fuel lines, the ability of the fuel to lubricate the injector pump and injectors adequately, and its ignition qualities (ignition delay, cetane number).

Inline mechanical injector pumps generally tolerate poor-quality or bio-fuels better than distributor-type pumps. Also, indirect injection engines generally run more satisfactorily on bio-fuels than direct injection engines. This is partly because an indirect injection engine has a much greater 'swirl' effect, improving vapourisation and combustion of fuel, and because (in the case of vegetable oil-type fuels) lipid depositions can condense on the cylinder walls of a direct-injection engine if combustion temperatures are too low (such as starting the engine from cold).

It is often reported that Diesel designed his engine to run on peanut oil. Diesel stated in his published papers, "at the Paris Exhibition in 1900 (*Exposition Universelle*) there was shown by the Otto Company a small diesel engine, which, at the request of the French Government ran on Arachide (earth-nut or pea-nut) oil, and worked so smoothly that only a few people were aware of it. The engine was constructed for using mineral oil, and was then worked on vegetable oil without any alterations being made. The French Government at the time thought of testing the applicability to power production of the Arachide, or earth-nut, which grows in considerable quantities in their African colonies, and can easily be cultivated there." Diesel himself later conducted related tests and appeared supportive of the idea.

Most large marine diesels (sometimes called *cathedral engines* due to their size) run on heavy fuel oil (sometimes called "bunker oil"), which is a thick, viscous and almost flameproof fuel which is very safe to store and cheap to buy in bulk as it is a waste product from the petroleum refining industry. The fuel must be heated to thin it out (often by the exhaust header) and is often passed through multiple injection stages to vapourise it.

Fuel and Fluid Characteristics

Diesel engines can operate on a variety of different fuels, depending on configuration, though the eponymous diesel fuel derived from crude oil is most common. The engines can work with the full spectrum of crude oil distillates, from natural gas, alcohols, petrol, wood gas to the *fuel oils* from diesel oil to residual fuels.

The type of fuel used is a combination of service requirements, and fuel costs. Good-quality diesel fuel can be synthesised from vegetable oil and alcohol. Diesel fuel can be made from coal or other carbon base using the Fischer-Tropsch process. Biodiesel is growing in popularity since it can frequently be used in unmodified engines, though production remains limited. Recently, biodiesel from coconut, which can produce a very promising coco methyl ester (CME), has characteristics which enhance lubricity and combustion giving a regular diesel engine without any modification more power, less particulate matter or black smoke, and smoother engine performance. The Philippines pioneers in the research on Coconut based CME with the help of German and American scientists. Petroleum-derived diesel is often called *petrodiesel* if there is need to distinguish the source of the fuel.

Pure plant oils are increasingly being used as a fuel for cars, trucks and remote combined heat and power generation especially in Germany where hundreds of decentralised small- and medium-sized oil presses cold press oilseed, mainly rapeseed, for fuel. There is a Deutsches Institut für Normung fuel standard for rapeseed oil fuel.

Residual fuels are the "dregs" of the distillation process and are a thicker, heavier oil, or oil with higher viscosity, which are so thick that they are not readily pumpable unless heated. Residual fuel oils are cheaper than clean, refined diesel oil, although they are dirtier. Their main considerations are for use in ships and very large generation sets, due to the cost of the large volume of fuel consumed, frequently amounting to many tonnes per hour. The poorly refined biofuels straight vegetable oil (SVO) and waste vegetable oil (WVO) can fall into this category, but can be viable fuels on non common rail or TDI PD diesels with the simple conversion of fuel heating to 80 to 100 degrees Celsius to reduce viscosity, and adequate filtration to OEM standards. Engines using these heavy oils have to start and shut down on standard diesel fuel, as these fuels will not flow through fuel lines at low temperatures. Moving beyond that, use of low-grade fuels can lead to serious maintenance problems because of their high sulphur and lower lubrication properties. Most diesel engines that power ships like supertankers are built so that the engine can safely use low-grade fuels due to their separate cylinder and crankcase lubrication.

Normal diesel fuel is more difficult to ignite and slower in developing fire than petrol because of its higher flash point, but once burning, a diesel fire can be fierce.

Fuel contaminants such as dirt and water are often more problematic in diesel engines than in petrol engines. Water can cause serious damage, due to corrosion, to the injection pump and injectors; and dirt, even very fine particulate matter, can damage the injection pumps due to the close tolerances that the pumps are machined to. All diesel engines will have a fuel filter (usually much finer than a filter on a petrol engine), and a water trap. The water trap (which is sometimes part of the fuel filter) often has a float connected to a warning light, which warns when there is too much water in the trap, and must be drained before damage to the engine can result. The fuel filter must be replaced much more often on a diesel engine than on a petrol engine, changing the fuel filter every 2-4 oil changes is not uncommon for some vehicles.

Safety—Fuel Flammability

Diesel fuel has low flammability, leading to a low risk of fire caused by fuel in a vehicle equipped with a diesel engine. In yachts diesels are used because petrol engines generate combustible vapours, which can accumulate in the bottom of the vessel, sometimes causing explosions. Therefore ventilation systems on petrol powered vessels are required. The United States Army and NATO use only diesel engines and turbines because of fire hazard. Although neither gasoline nor diesel is explosive in liquid form, both can create an explosive air/

vapour mix under the right conditions. However, diesel fuel is less prone due to its lower vapour pressure, which is an indication of evapouration rate. The Material Safety Data Sheet for ultra-low sulfur diesel fuel indicates a vapour explosion hazard for diesel indoors, outdoors, or in sewers. US Army gasoline-engined tanks during World War II were nicknamed Ronsons, because of their greater likelihood of catching fire when damaged by enemy fire. (Although tank fires were usually caused by detonation of the ammunition rather than fuel.)

Diesel Applications

The characteristics of diesel have different advantages for different applications.

Passenger Cars

Diesel engines have long been popular in bigger cars and this is spreading to smaller cars. Diesel engines tend to be more economical at regular driving speeds and are much better at city speeds. Their reliability and life-span tend to be better (as detailed). Some 40% or more of all cars sold in Europe are diesel-powered where they are considered a low CO_2 option. Mercedes-Benz in conjunction with Robert Bosch GmbH produced diesel-powered passenger cars starting in 1936 and very large numbers are used all over the world (often as "Grande Taxis" in the Third World).

Railroad Rolling Stock

Diesel engines have eclipsed steam engines as the prime mover on all non-electrified railroads in the industrialised world. The first diesel locomotives appeared in the early 20th century, and diesel multiple units soon after. While electric locomotives have now replaced the diesel locomotive almost completely on passenger traffic in Europe and Asia, diesel is still today very popular for cargo-hauling freight trains and on tracks where electrification is not feasible. Most modern diesel locomotives are actually diesel-electric locomotives: the diesel engine is used to power an electric generator that in turn powers electric traction engines with no mechanical connection between diesel engine and traction.

Other Transport Uses

Larger transport applications (trucks, buses, etc.) also benefit from the diesel's reliability and high torque output. Diesel displaced paraffin (or tractor vapourising oil, TVO) in most parts of the world by the end of the 1950s with the U.S. following some 20 years later.

- Aircraft
- Marine
- Motorcycles.

In merchant ships and boats, the same advantages apply with the relative safety of diesel fuel an additional benefit. The German pocket battleships were the largest diesel warships, but the German torpedo-boats known as E-boats (*Schnellboot*) of the Second World War were also diesel craft. Conventional submarines have used them since before the First World War, relying on the almost total absence of carbon monoxide in the exhaust. American World War II diesel-electric submarines operated on two-stroke cycle as opposed to the four-stroke cycle that other navies used.

Military Fuel Standardisation

NATO has a single vehicle fuel policy and has selected diesel for this purpose. The use of a single fuel simplifies wartime logistics. NATO and the United States Marine Corps have even been developing a diesel military motorcycle based on a Kawasaki off road motorcycle, with a purpose designed naturally aspirated direct injection diesel at Cranfield University in England, to be produced in the USA, because motorcycles were the last remaining gasoline-powered vehicle in their inventory. Before this, a few civilian motorcycles had been built using adapted stationary diesel engines, but the weight and cost disadvantages generally outweighed the efficiency gains.

Non-transport Uses

Diesel engines are also used to power permanent, portable, and backup generators, irrigation pumps, corn grinders, and coffee de-pulpers.

Engine Speeds

Within the diesel engine industry, engines are often categorised by their rotational speeds into three unofficial groups:

- High-speed engines,
- medium-speed engines, and
- slow-speed engines.

High- and medium-speed engines are predominantly four-stroke engines. Medium-speed engines are physically larger than high-speed engines and can burn lower-grade (slower-burning) fuel than high-speed engines. Slow-speed engines are predominantly large two-stroke crosshead engines, hence very different from high- and medium-speed engines. Due to the lower rotational speed of slow- and medium-speed engines, there is more time for combustion during the power stroke of the cycle, allowing the use of slower-burning fuels than high-speed engines.

High-speed Engines

High-speed (approximately 1,000 rpm and greater) engines are used to power trucks (lorries), buses, tractors, cars, yachts, compressors, pumps and small electrical generators. As of 2008, most high-speed engines have direct injection. Many modern engines, particularly in on-highway applications, have common rail direct injection, which is cleaner burning.

Medium-speed Engines

Medium-speed engines are used in large electrical generators, ship propulsion and mechanical drive applications such as large compressors or pumps. Medium speed diesel engines operate on either diesel fuel or heavy fuel oil by direct injection in the same manner as low-speed engines.

Engines used in electrical generators run at approximately 300 to 1000 rpm and are optimised to run at a set synchronous speed depending on the generation frequency (50 or 60 hertz) and provide a rapid response to load changes. Typical synchronous speeds for modern medium-speed engines are 500/514 rpm (50/60 Hz), 600 rpm (both 50 and 60 Hz), 720/750 rpm, and 900/1000 rpm.

As of 2009, the largest medium-speed engines in current production have outputs up to approximately 20 MW (27,000 hp). and are supplied by companies like MAN B&W, Wärtsilä, and Rolls-Royce. Most medium-speed engines produced are four-stroke machines, however there are some two-stroke medium-speed engines such as by EMD (Electro-Motive Diesel), and the Fairbanks Morse OP (Opposed-piston engine) type.

Typical cylinder bore size for medium-speed engines ranges from 20 cm to 50 cm, and engine configurations typically are offered ranging from in-line 4-cylinder units to V-configuration 20-cylinder units. Most larger medium-speed engines are started with compressed air direct on pistons, using an air distributor, as opposed to a pneumatic starting motor acting on the flywheel, which tends to be used for smaller engines. There is no definitive engine size cut-off point for this.

It should also be noted that most major manufacturers of medium-speed engines make natural gas-fuelled versions of their diesel engines, which in fact operate on the Otto cycle, and require spark ignition, typically provided with a spark plug. There are also dual (diesel/natural gas/coal gas) fuel versions of medium and low speed diesel engines using a lean fuel air mixture and a small injection of diesel fuel (so-called "pilot fuel") for ignition. In case of a gas supply failure or maximum power demand these engines will instantly switch back to full diesel fuel operation.

Low-speed Engines

Figure 15: The MAN B&W 5S50MC 5-cylinder, 2-stroke, low-speed marine diesel engine. This particular engine is found aboard a 29,000 tonne chemical carrier.

Also known as *slow-speed*, or traditionally *oil engines*, the largest diesel engines are primarily used to power ships, although there are a few land-based power generation units as well. These extremely large two-stroke engines have power outputs up to approximately 85 MW (114,000 hp), operate in the range from approximately 60 to 200 rpm and are up to 15 m (50 ft) tall, and can weigh over 2,000 short tons (1,800 t). They typically use direct injection running on cheap low-grade heavy fuel, also known as *Bunker C* fuel, which requires heating in the ship for tanking and before injection due to the fuel's high viscosity.

The heat for fuel heating is often provided by waste heat recovery boilers located in the exhaust ducting of the engine, which produce the steam required for fuel heating. Provided the heavy fuel system is kept warm and circulating, engines can be started and stopped on heavy fuel.

Large and medium marine engines are started with compressed air directly applied to the pistons. Air is applied to cylinders to start the engine forwards or backwards because they are normally directly connected to the propeller without clutch or gearbox, and to provide reverse propulsion either the engine must be run backwards or the ship will utilise an adjustable propeller. At least three cylinders are required with two-stroke engines and at least six cylinders with four-stroke engines to provide torque every 120 degrees.

Companies such as MAN B&W Diesel, (formerly Burmeister & Wain) and Wärtsilä (which acquired Sulzer Diesel) design such large low-speed engines. They are unusually narrow and tall due to the addition of a crosshead bearing. As of 2007, the 14-cylinder Wärtsilä-Sulzer 14RTFLEX96-C turbocharged two-stroke diesel engine built by Wärtsilä licensee Doosan in Korea is the most powerful diesel engine put into service, with a cylinder bore of 960 mm (37.8 in) delivering 114,800 hp (85.6 MW). It was put into service in September 2006, aboard the world's largest container ship *Emma Maersk* which belongs to the A.P. Moller-Maersk Group. Typical bore size for low-speed engines ranges from approximately 35 to 98 cm (14 to 39 in). As of 2008, all produced low-speed engines with crosshead bearings are in-line configurations; no Vee versions have been produced.

Supercharging and Turbocharging

Most diesels are now turbocharged and some are both turbo charged and supercharged. Because diesels do not have fuel in the cylinder before combustion is initiated, more than one bar (100 kPa) of air can be loaded in the cylinder without preignition. A turbocharged engine can produce significantly more power than a naturally aspirated engine of the same configuration, as having more air in the cylinders allows more fuel to be burned and thus more power to be produced. A supercharger is powered mechanically by the engine's crankshaft, while a turbocharger is powered by the engine exhaust, not requiring any mechanical power. Turbocharging can improve the fuel economy of diesel engines by recovering waste heat from the exhaust, increasing the excess air factor, and increasing the ratio of engine output to friction losses.

A two-stroke engine does not have a discrete exhaust and intake stroke and thus is incapable of self-aspiration. Therefore all two-stroke engines must be fitted with a blower to charge the cylinders with air and assist in dispersing exhaust gases, a process referred to as scavenging. In some cases, the engine may also be fitted with a turbocharger, whose output is directed into the blower inlet. A few designs employ a hybrid turbocharger for scavenging and charging the cylinders, which device is mechanically driven at cranking and low speeds to act as a blower.

As turbocharged or supercharged engines produce more power for a given engine size as compared to naturally aspirated engines, attention must be paid to the mechanical design of components, lubrication, and cooling to handle the power. Pistons are usually cooled with lubrication oil sprayed on the bottom of the piston. Large engines may use water, sea water, or oil supplied through telescoping pipes attached to the crosshead.

Current and Future Developments

As of 2008, many common rail and unit injection systems already employ new injectors using stacked piezoelectric wafers in lieu of a solenoid, giving finer control of the injection event.

Variable geometry turbochargers have flexible vanes, which move and let more air into the engine depending on load. This technology increases both performance and fuel economy. Boost lag is reduced as turbo impeller inertia is compensated for.

Accelerometre pilot control (APC) uses an accelerometre to provide feedback on the engine's level of noise and vibration and thus instruct the ECU to inject the minimum amount of fuel that will produce quiet combustion and still provide the required power (especially while idling).

The next generation of common rail diesels is expected to use variable injection geometry, which allows the amount of fuel injected to be varied over a wider range, and variable valve timing similar to that on petrol engines. Particularly in the United States, coming tougher emissions regulations present a considerable challenge to diesel engine manufacturers. Ford's Hy Trans Project has developed a system which starts the ignition in 400 ms, saving a significant amount of fuel on city routes, and there are other methods to achieve even more efficient combustion, such as homogeneous charge compression ignition, being studied.

Atkinson Cycle

The Atkinson cycle engine is a type of internal combustion engine invented by James Atkinson in 1882. The Atkinson cycle is designed to provide efficiency at the expense of power density, and is used in some modern hybrid electric applications.

Design

The original Atkinson cycle piston engine allowed the intake, compression, power, and exhaust strokes of the four-stroke cycle to occur in a single turn of the crankshaft and was designed to avoid infringing certain patents covering Otto cycle engines. Due to the unique crankshaft design of the Atkinson, its expansion ratio can differ from its compression ratio and, with a power stroke longer than its compression stroke, the engine can achieve greater thermal efficiency than a traditional piston engine. While Atkinson's original design is no more than an historical curiosity, many modern engines use unconventional valve timing to produce the effect of a shorter compression stroke/longer power stroke, thus realising the fuel economy improvements the Atkinson cycle can provide.

Ideal Thermodynamic Cycle

The ideal Atkinson cycle consists of following operations:

- 1-2 Isentropic or reversible adiabatic compression
- 2-3 Isochoric heating (Qp)
- 3-4 Isobaric heating (Qp')
- 4-5 Isentropic expansion
- 5-6 Isochoric cooling (Qo)
- 6-1 Isobaric cooling (Qo').

Modern Atkinson Cycle Engines

Recently Atkinson cycle has been used to describe a modified Otto cycle engine in which the intake valve is held open longer than normal to allow a reverse flow of intake air into the intake manifold. The effective compression ratio is reduced (for a time the air is escaping the cylinder freely rather than being compressed) but the expansion ratio is unchanged. This means the compression ratio is smaller than the expansion ratio. Heat gained from burning fuel increases the pressure, thereby forcing the piston to move, expanding the air volume beyond the volume when compression began.

The goal of the modern Atkinson cycle is to allow the pressure in the combustion chamber at the end of the power stroke to be equal to atmospheric pressure; when this occurs, all the available energy has been obtained from the combustion process. For any given portion of air, the greater expansion ratio allows more energy to be converted from heat to useful mechanical energy meaning the engine is more efficient.

The disadvantage of the four-stroke Atkinson cycle engine versus the more common Otto cycle engine is reduced power density. Due to a smaller portion of the compression stroke being devoted to compressing the intake air, an Atkinson cycle engine does not take in as much air as would a similarly designed and sized Otto cycle engine. Four-stroke engines of this type with this same type of intake valve motion but with a supercharger to make up for the loss of power density are known as Miller cycle engines.

Rotary Atkinson Cycle Engine

The Atkinson cycle can be used in a rotary engine. In this configuration an increase in both power and efficiency can be achieved when compared to the Otto cycle. This type of engine retains the one power phase per revolution, together with the different compression and expansion volumes of the original Atkinson cycle. Exhaust gases are expelled from the engine by compressed-air scavenging. This modification of the Atkinson cycle allows for the use of alternative fuels like diesel and hydrogen. Disadvantages of this design include the requirement that rotor tips seal very tightly on the outer housing wall and the mechanical losses suffered through friction between rapidly oscillating parts of irregular shape.

Vehicles Using Atkinson Cycle Engines

While a modified Otto cycle engine using the Atkinson cycle provides good fuel economy, it is at the expense of a lower power-per-displacement as compared to a traditional four-stroke engine. If demand for more power is intermittent, the power of the engine can be supplemented by an electric motor during times when more power is needed. This forms the basis of an Atkinson cycle-based hybrid electric drivetrain. These electric motors can be used independently of, or in combination with, the Atkinson cycle engine, to provide the most efficient means of producing the desired power.

Several production vehicles use Atkinson cycle engines:

- Chevrolet Tahoe Hybrid electric (four-wheel drive) with a compression ratio of 10.8:1
- Ford Escape/Mercury Mariner/Mazda Tribute electric (front- and four-wheel drive) with a compression ratio of 12.4:1
- Ford Fusion Hybrid/Mercury Milan Hybrid/Lincoln MKZ Hybrid electric (front-wheel drive) with a compression ratio of 12.3:1
- Hyundai Sonata Hybrid (front-wheel drive)

- Infiniti M35h Hybrid (rear-wheel drive)
- Kia Optima Hybrid (front-wheel drive)
- Lexus CT200H (front-wheel drive)
- Lexus HS250h (front-wheel drive)
- Lexus RX 450h hybrid electric (front-wheel drive)
- Mercedes ML450 Hybrid (four-wheel drive) electric
- Mercedes S400 Blue Hybrid (rear-wheel drive) electric
- Toyota Highlander Hybrid (2011 and newer)
- Toyota Prius hybrid electric (front-wheel drive) with a (purely geometric) compression ratio of 13.0:1
- Toyota Camry Hybrid electric (front-wheel drive) with a compression ratio of 12.5:1
- Lexus GS450h hybrid electric (Rear-Wheel drive) with a compression ratio of 13.0:1

Miller Cycle

In engineering, the Miller cycle is a combustion process used in a type of four-stroke internal combustion engine. The Miller cycle was patented by Ralph Miller, an American engineer, in the 1940s.

Overview

This type of engine was first used in ships and stationary power-generating plants, but was adapted by Mazda for their *KJ-ZEM* V6, used in the Millenia sedan, and in their Eunos 800 sedan (Australia) luxury cars. More recently, Subaru has combined a Miller cycle flat-4 with a hybrid driveline for their concept "Turbo Parallel Hybrid" car, known as the Subaru B5-TPH. A traditional Otto cycle engine uses four "strokes", of which two can be considered "high power" — the compression stroke (high power consumption) and power stroke (high power production). Much of the internal power loss of an engine is due to the energy needed to compress the charge during the compression stroke, so systems that reduce this power consumption can lead to greater efficiency.

In the Miller cycle, the intake valve is left open longer than it would be in an Otto cycle engine. In effect, the compression stroke is two discrete cycles: the initial portion when the intake valve is open and final portion when the intake valve is closed. This two-stage intake stroke creates the so called "fifth" stroke that the Miller cycle introduces. As the piston initially moves upwards in what is traditionally the compression stroke, the charge is partially expelled back out the still-open intake valve. Typically this loss of charge air would result in a loss of power. However, in the Miller cycle, this is compensated for by the use of a supercharger. The supercharger typically will need to be of the positive displacement (Roots or Screw) type due to its ability to produce boost at relatively low engine speeds. Otherwise, low-rpm torque will suffer.

A key aspect of the Miller cycle is that the compression stroke actually starts only after the piston has pushed out this "extra" charge and the intake valve closes. This happens at around 20% to 30% into the compression stroke. In other words, the actual compression occurs in the latter 70% to 80% of the compression stroke.

In a typical spark ignition engine, the Miller cycle yields an additional benefit. The intake air is first compressed by the supercharger and then cooled by an intercooler. This lower intake charge temperature, combined with the lower compression of the intake stroke, yields a lower final charge temperature than would be obtained by simply increasing the compression of the piston. This allows ignition timing to be advanced beyond what is normally allowed before the onset of detonation, thus increasing the overall efficiency still further. An additional advantage of the lower final charge temperature is that the emission of NOx in diesel engines is decreased, which is an important design parameter in large diesel engines on board ships and power plants.

Efficiency is increased by raising the compression ratio. In a typical gasoline engine, the compression ratio is limited due to self-ignition (detonation) of the compressed, and therefore hot, air/fuel mixture. Due to the reduced compression stroke of a Miller cycle engine, a higher overall cylinder pressure (supercharger pressure plus mechanical compression) is possible, and therefore a Miller cycle engine has better efficiency. The benefits of utilising positive displacement superchargers come with a cost. 15% to 20% of the power generated by a supercharged engine is usually required to do the work of driving the supercharger, which compresses the intake charge (also known as boost).

A similar delayed-valve closing method is used in some modern versions of Atkinson cycle engines, but without the supercharging. These engines are generally found on hybrid electric vehicles, where efficiency is the goal, and the power lost compared to the Miller cycle is made up through the use of electric motors.

Chapter 5

Pistonless Rotary Engine

A pistonless rotary engine is an internal combustion engine that does not use pistons in the way a reciprocating engine does, but instead uses one or more rotors, sometimes called rotary pistons. An example of a pistonless rotary engine is the Wankel engine.

The term *rotary combustion engine* has been suggested as an alternative name for these engines to distinguish them from early (generally up to the early 1920s) aircraft engines and motorcycle engines also known as *rotary engines*. However, both continue to be called *rotary engines* and only the context determines which type is meant.

Pistonless Rotary Engines

The basic concept of a (pistonless) rotary engine avoids the reciprocating motion of the piston with its inherent vibration and rotational-speed-related mechanical stress. As of 2006 the Wankel engine is the only successful pistonless rotary engine, but many similar concepts have been proposed and are under various stages of development. Examples of rotary engines include:

Production stage:

- The Wankel engine.

Development stage:

- The original Atkinson cycle engine
- The Engineair engine
- The Liquidpiston engine
- The Sarich orbital engine
- The Quasiturbine
- The Ramgen Integrated Supersonic Component Engine
- The Rand cam engine

- The RKM engine (Rotations Kolben Maschine)
- The Trochilic engine
- The Wave disk engine.

Conceptual stage:

- The Gerotor engine.

Wankel Engine

The Wankel engine is a type of internal combustion engine using an eccentric rotary design to convert pressure into a rotating motion instead of using reciprocating pistons. Its four-stroke cycle takes place in a space between the inside of an oval-like epitrochoid-shaped housing and a rotor that is similar in shape to a Reuleaux triangle but with sides that are somewhat flatter.

The very compact Wankel engine delivers smooth high-rpm power. It is commonly called a rotary engine, though this name applies also to other completely different designs. It is the only internal combustion engine invented in the twentieth century to go into production.

The engine was invented by German engineer Felix Wankel. He received his first patent for the engine in 1929, began development in the early 1950s at NSU, completing a working prototype in 1957. NSU then licensed the concept to companies around the world, which have continued to improve the design.

Thanks to their compact design, Wankel rotary engines have been installed in a variety of vehicles and devices including automobiles, motorcycles, racers, aircraft, go-karts, jet skis, snowmobiles, chain saws, and auxiliary power units. Perhaps the greatest proponent of the Wankel engine has been the Japanese company Mazda.

History

In 1951, the German engineer Felix Wankel began development of the engine at NSU Motorenwerke AG, where he first conceived his rotary engine in 1954 (DKM 54, *Drehkolbenmotor*). The so-called KKM 57 was constructed by NSU engineer Hanns Dieter Paschke in 1957 without the knowledge of Felix Wankel, who remarked "*you have turned my race horse into a plow mare*". The first working prototype DKM 54 was running on February 1, 1957 at the NSU research and development department *Versuchsabteilung TX*. It produced 21 horsepower; unlike modern Wankel engines, both the rotor and the housing rotated.

Considerable effort went into designing rotary engines in the 1950s and 1960s. They were of particular interest because they were smooth and quiet running, and because of the reliability resulting from their simplicity. An early problem of buildup of cracks in the epitrochoid surface was solved by installing the spark plugs in a separate metal piece instead of screwing them directly into the block.. A later alternative solution to spark plug boss cooling was provided by variable coolant velocity scheme for water-cooled rotaries which has had widespread use and was patented by Curtiss-Wright, with the last-listed for better air-cooled engine spark plug boss cooling. These approaches did not require a high conductivity copper insert but did not preclude the use.

Among the manufacturers signing licensing agreements to develop Wankel engines were Alfa Romeo, American Motors, Citroen, Ford, General Motors, Mercedes-Benz, Nissan, Porsche, Rolls-Royce, Suzuki, and Toyota. In the United States, in 1959 under license from NSU, Curtiss-Wright pioneered improvements in the basic engine design. In Britain, in the 1960s, Rolls Royce Motor Car Division pioneered a two-stage diesel version of the Wankel engine.

Also in Britain, Norton Motorcycles developed a Wankel rotary engine for motorcycles, based on the Sachs air-cooled Wankel that powered the DKW/Hercules W-2000 motorcycle, which was included in their Commander and F1; Suzuki also made a production motorcycle with a Wankel engine, the RE-5, where they used ferrotic alloy apex seals and an NSU rotor in a successful attempt to prolong the engine's life. In 1971 and 1972 Arctic Cat produced snowmobiles powered by 303 cc Wankel rotary engines manufactured by Sachs in Germany. Deere & Company designed a version that was capable of using a variety of fuels. The design was proposed as the power source for United States Marine Corps combat vehicles and other equipment in the late 1980s.

Mazda and NSU signed a study contract to develop the Wankel engine in 1961 and competed to bring the first Wankel powered automobile to market. Although Mazda produced an experimental Wankel that year, NSU was first with a Wankel automobile on sale, the sporty NSU Spider in 1964; Mazda countered with a display of two and four rotor Wankel engines at that year's Tokyo Motor Show. In 1967, NSU began production of a Wankel-engined luxury car, the Ro 80. However, problems with apex seal wear led to frequent engine failure, which led to large warranty costs for NSU, and curtailed further Wankel engine development.

Figure 1: Mazda's first Wankel engine, at the Mazda Museum in Hiroshima, Japan

Mazda, however, claimed to have solved the apex seal problem, and was able to run test engines at high speed for 300 hours without failure. After years of development, Mazda's first Wankel engine car was the 1967 Cosmo 110S. The company followed with a number of Wankel ("rotary" in the company's terminology) vehicles, including a bus and a pickup truck. Customers often cited the cars' smoothness of operation. However, Mazda chose a method to comply with hydrocarbon emission standards that, while less expensive to produce, increased fuel consumption, just before a sharp rise in fuel prices. Mazda later abandoned the Wankel in most of their automotive designs, but continued using it in their RX-7 sports car until August 2002 (RX-7 importation for Canada ceased with only the 1993 year being sold. The USA ended with the 1994 model year with remaining unsold stock being carried over as the '1995' year.).

The company normally used two-rotor designs, but the 1991 Eunos Cosmo used a twin-turbo three-rotor engine. In 2003, Mazda introduced the Renesis engine with the RX-8. The Renesis engine relocated the ports for exhaust and intake from the periphery of the rotary housing to the sides, allowing for larger overall ports, better airflow, and further power gains. Early Wankel engines had also side intake and exhaust ports, but the concept was abandoned because of carbon buildup in ports and side of rotor. The Renesis engine solved the problem by using a keystone scraper side seal. The Renesis is capable of delivering 238 hp (177 kW) with better fuel economy, reliability, and environmental friendliness than previous Mazda rotary engines, all from a nominal 1.3 L displacement.

In 1961, the Soviet research organisation of NATI, NAMI and VNII motoprom started experimental development, and created experimental engines with different technologies.

Soviet automobile manufacturer AvtoVAZ also experimented with the use of Wankel engines in cars but without the benefit of a license. In 1974 they created a special engine design bureau, which in 1978 designed an engine designated as VAZ-311. In 1980, the company started delivering Wankel-powered VAZ-2106s (VAZ-411 engine with two-rotors) and Ladas, mostly to security services, of which about 200 were made. The next models were the VAZ-4132 and VAZ-415. Aviadvigatel, the Soviet aircraft engine design bureau, is known to have produced Wankel engines with electronic injection for aircraft and helicopters, though little specific information has surfaced.

Although many manufacturers licensed the design, including Citroën with their M35 and GS Birotor, using engines produced by Comotor, General Motors, which seems to have concluded that the Wankel engine was slightly more expensive to build than an equivalent reciprocating engine, although claiming having solved the fuel economy issue, but failed in obtaining acceptable exhaust emissions, and Mercedes-Benz which used it for their C111 concept car, only Mazda has produced Wankel engines in large numbers.

American Motors (AMC) was so convinced "...that the rotary engine will play an important role as a powerplant for cars and trucks of the future...", according to Chairman Roy D. Chapin Jr., that the smallest U.S. automaker signed an agreement in February 1973, after a year's negotiations, to build Wankels for both passenger cars and Jeeps, as well as the right to sell any rotary engines it produces to other companies.

It even designed the unique Pacer around the engine, even though by then, AMC had decided to buy the Wankel engines from GM instead of building them itself. However, GM's engines had not reached production when the Pacer was to hit the showrooms. Part of the demise of this feature was the 1973 oil crisis with rising fuel prices, and also concerns about proposed US emission standards legislation. General Motors' Wankel did not comply with those emission standards, so in 1974 the company canceled its development, although GM claimed having solved the fuel consumption problem; unfortunately, they just published one SAE paper on the results of their research. This meant the Pacer had to be reconfigured to house AMC's venerable AMC Straight-6 engine with rear-wheel drive.

Design

In the Wankel engine, the four strokes of a typical Otto cycle occur in the space between a three-sided symmetric rotor and the inside of a housing. The expansion phase of the Wankel cycle is much longer than that of the Otto cycle. In the basic single-rotor Wankel engine, the oval-like epitrochoid-shaped housing surrounds a rotor which is triangular with bow-shaped flanks. The theoretical shape of the rotor between the fixed corners is the result of a minimisation of the volume of the geometric combustion chamber and a maximisation of the compression ratio, respectively. The symmetric curve connecting two arbitrary apexes of the rotor is maximised in the direction of the inner housing shape with the constraint that it not touch the housing at any angle of rotation (an arc is not a solution of this optimisation problem).

The central drive shaft, called the eccentric shaft or E-shaft, passes through the centre of the rotor and is supported by fixed bearings. The rotors ride on eccentrics (analogous to cranks) integral to the eccentric shaft (analogous to a crankshaft). The rotors both rotate around the eccentrics and make orbital revolutions around the eccentric shaft. Seals at the corners of the rotor seal against the periphery of the housing, dividing it into three moving combustion chambers. The rotation of each rotor on its own axis is caused and controlled by a pair of synchronizing gears A fixed gear mounted on one side of the rotor housing engages a ring gear attached to the rotor and ensures the rotor moves exactly 1/3 turn for each turn of the eccentric shaft. The power output of the engine is not transmitted through the synchronizing gears. The force of gas pressure on the rotor (to a first approximation) goes directly to the centre of the eccentric, part of the output shaft.

The best way to visualise the action of the engine in the animation at left is to look not at the rotor itself, but the cavity created between it and the housing. The Wankel engine is actually a variable-volume progressing-cavity system. Thus there are 3 cavities per housing, all repeating the same cycle. Note as well that points A and B on the rotor and e-shaft turn at different speed, point B moves 3 times faster than point A, so that one full orbit of the rotor equates to 3 turns of the e-shaft.

As the rotor rotates and orbitally revolves, each side of the rotor is brought closer to and then away from the wall of the housing, compressing and expanding the combustion chamber like the strokes of a piston in a reciprocating engine. The power vector of the combustion stage goes through the centre of the offset lobe.

While a four-stroke piston engine makes one combustion stroke per cylinder for every two rotations of the crankshaft (that is, one-half power stroke per crankshaft rotation per cylinder), each combustion chamber in the Wankel generates one combustion stroke per each driveshaft rotation, i.e. one power stroke per rotor orbital revolution and three power strokes per rotor rotation. Thus, power output of a Wankel engine is generally higher than that of a four-stroke piston engine of similar engine displacement in a similar state of tune; and higher than that of a four-stroke piston engine of similar physical dimensions and weight.

Wankel engines also generally have a much higher redline than a reciprocating engine of similar power output. This is in part because the smoothness inherent in circular motion, but especially because they do not have highly stressed parts such as a crankshaft or connecting rods. Eccentric shafts do not have the stress-raising internal corners of crankshafts. The redline of a rotary engine is limited by wear of the synchronizing gears. Hardened steel gears are used for extended operation above 7000 or 8000 rpm. Mazda Wankel engines in auto racing are operated above 10,000 rpm. In aircraft they are used conservatively, up to 6500 or 7500 rpm. However, as gas pressure participates in seal efficiency, running a Wankel engine at high rpm under no load conditions can destroy the engine.

National agencies that tax automobiles according to displacement and regulatory bodies in automobile racing variously consider the Wankel engine to be equivalent to a four-stroke engine of 1.5 to 2 times the displacement; some racing series ban it altogether.

Engineering

Felix Wankel managed to overcome most of the problems that made previous rotary engines fail by developing a configuration with vane seals that had a tip radius equal to the amount of "oversize" of the rotor housing form, as compared to the theoretical epitrochoid, to minimise radial apex seal motion plus introducing a cylindrical gas-loaded apex pin which abutted all sealing elements to seal around the 3 planes at each rotor apex.

Rotary engines have a thermodynamic problem not found in reciprocating four-stroke engines in that their "cylinder block" operates at steady state, with intake, compression, combustion, and exhaust occurring at fixed housing locations for all "cylinders". In contrast, reciprocating engines perform these four strokes in one chamber, so that extremes of "freezing" intake and "flaming" exhaust are averaged and shielded by a boundary layer from overheating working parts.

The boundary layer shields and the oil film act as thermal insulation, leading to a low temperature of the lubricating film (max. ~200 °C/400 °F) on a water-cooled Wankel engine. This gives a more constant surface temperature. The temperature around the spark plug is about the same as the temperature in the combustion chamber of a reciprocating engine. With circumferential or axial flow cooling, the temperature difference remains tolerable.

Four-stroke reciprocating engines are less suitable for hydrogen. The hydrogen can misfire on hot parts like the exhaust valve and spark plugs. Another problem concerns the hydrogenate attack on the lubricating film in reciprocating engines. In a Wankel engine, this problem is circumvented by using a ceramic apex seal against a ceramic surface: there is no oil film to suffer hydrogenate attack. Since ceramic piston rings are not available as of 2009, the problem remains with the reciprocating engine. The piston shell must be lubricated and cooled with oil. This substantially increases the lubricating oil consumption in a four-stroke hydrogen engine.

Materials

Unlike a piston engine, where the cylinder is cooled by the incoming charge after being heated by combustion, Wankel rotor housings are constantly heated on one side and cooled on the other, leading to high local temperatures and unequal thermal expansion. While this places high demands on the materials used, the simplicity of the Wankel makes it easier to use alternative materials like exotic alloys and ceramics. With water cooling in a radial or axial flow direction, with the hot water from the hot bow heating the cold bow, the thermal expansion remains tolerable.

Sealing

Early engine designs had a high incidence of sealing loss, both between the rotor and the housing and also between the various pieces making up the housing. Also, in earlier model Wankel engines carbon particles could become trapped between the seal and the casing, jamming the engine and requiring a partial rebuild. It was common for very early Mazda engines to require rebuilding after 50,000 miles (80,000 km). Further sealing problems arise from the uneven thermal distribution within the housings causing distortion and loss of sealing and compression. This thermal distortion also causes uneven wear between the apex seal and the rotor housing, quite evident on higher mileage engines. The problem is exacerbated when the engine is stressed before reaching operating temperature. However, Mazda Wankel engines have solved these problems. Current engines have nearly 100 seal-related parts.

The problem of clearance for hot rotor apices passing between the axially closer side housings in the cooler intake lobe areas was dealt with by using an axial rotor pilot, radially inboard of the oils seals plus improved inertia oil cooling of the rotor interior.

Fuel Consumption and Emissions

Just as the shape of the Wankel combustion chamber is resistant to preignition and will run on lower-octane rating gasoline than a comparable piston engine, it also leads to relatively incomplete combustion of the air-fuel charge, with a larger amount of unburned hydrocarbons released into the exhaust. The exhaust is, however, relatively low in NOx emissions; this allowed Mazda to meet the United States Clean Air Act of 1970 in 1973 with a simple and inexpensive 'thermal reactor' (an enlarged open chamber in the exhaust manifold) by paradoxically enriching the air-fuel ratio to the point where the unburned hydrocarbons (HC) in the exhaust would support complete combustion in the thermal reactor; while piston-engine cars required expensive catalytic converters to deal with both unburned hydrocarbons and NOx emissions.

This raised fuel consumption, however (already a weak point for the Wankel engine), at the same time that the oil crisis of 1973 raised the price of gasoline. Mazda was able to improve the fuel efficiency of the thermal reactor system by 40% by the time of introduction of the RX-7 in 1978, but eventually shifted to the catalytic converter system. According to the Curtiss-Wright research, the extreme that controls the amount of unburned HC in the exhaust is the rotor surface temperature, higher temperatures producing less HC. They showed also that the rotor can be widened. Quenching is the dominant source of HC at high speeds, and leakage at low speeds. Automobile Wankel rotary engines are high speed engines, however, it was shown that an early opening of the intake port, and long intake ducts can provide the required amount of torque at low RPM, and thus elasticity. The shape and positioning of rotor recess-combustion chamber-influences emissions and fuel use, the MDR being chosen as a compromise.

In Mazda's RX-8 with the Renesis engine, fuel consumption is now within normal limits while passing California State emissions requirements, including California's Low Emissions Vehicle (LEV) standards. The exhaust ports, which in earlier Mazda rotaries were located in the rotor housings, were moved to the sides of the combustion chamber. This approach allowed Mazda to eliminate overlap between intake and exhaust port openings, while simultaneously increasing exhaust port area. The side port trapped the unburned fuel in the chamber, decreased the oil consumption, and improved the combustion stability in the low-speed and light load range. The HC emissions from the side exhaust port Wankel engine are 35–50% less than those from the peripheral exhaust port Wankel engine.

Advantages

Wankel engines are considerably lighter, simpler, and contain far fewer moving parts than piston engines of equivalent power output. For instance, because valving is accomplished by simple ports cut into the walls of the rotor or side housings, they have no valves or complex valve trains; in addition, since the rotor rides directly on a large bearing on the output shaft, there are no connecting rods and there is no crankshaft. The elimination of reciprocating mass and the elimination of the most highly stressed and failure prone parts of piston engines gives the Wankel engine high reliability, a smoother flow of power, and a high power-to-weight ratio.

The surface/volume-ratio problem is so complex that one cannot make a direct comparison between a reciprocating piston engine and a Wankel engine in terms of the surface/volume-ratio. The flow velocity and the heat losses behave quite differently. Surface temperatures behave absolutely differently; the film of oil in the Wankel engine acts as insulation. Engines with a higher compression ratio have a worse surface/volume-ratio.

The surface/volume-ratio of a Diesel engine is much worse than a gasoline engine, but Diesel engines are well known for a higher efficiency factor than gasoline engines. Thus, engines with equal power should be compared: a naturally aspirated 1.3-litre Wankel engine with a naturally aspirated 1.3-litre four-stroke reciprocating piston engine with equal power. But such a four-stroke engine is not possible and needs twice the displacement for the same power as a Wankel engine. The extra or "empty" stroke(s) should not be ignored, as a 4-stroke cylinder produces a power stroke only every other rotation of the crankshaft. In actuality, this doubles the real surface/

volume-ratio for the four-stroke reciprocating piston engine and the demand of displacement. The Wankel, therefore, has higher volumetric efficiency and a lower pumping loss through the absence of choking valves. Because of the quasi-overlap of the power strokes that cause the smoothness of the engine and the avoidance of the 4-stroke cycle in a reciprocating engine, the Wankel engine is very quick to react to throttle changes and is able to quickly deliver a surge of power when the demand arises, especially at higher rpm. This difference is more pronounced when compared to four-cylinder reciprocating engines and less pronounced when compared to higher cylinder counts.

In addition to the removal of internal reciprocating stresses by virtue of the complete removal of reciprocating internal parts typically found in a piston engine, the Wankel engine is constructed with an iron rotor within a housing made of aluminium, which has a greater coefficient of thermal expansion.

This ensures that even a severely overheated Wankel engine cannot seize, as would be likely to occur in an overheated piston engine. This is a substantial safety benefit of use in aircraft. In addition, valves and valve trains that do not exist cannot burn out, jam, break, or malfunction in any way, again increasing safety.

A further advantage of the Wankel engine for use in aircraft is the fact that a Wankel engine generally has a smaller frontal area than a piston engine of equivalent power, allowing a more aerodynamic nose to be designed around it. The simplicity of design and smaller size of the Wankel engine also allows for savings in construction costs, compared to piston engines of comparable power output.

Wankel engines that operate within their original design parameters are almost immune to catastrophic failure. A Wankel engine that loses compression, cooling or oil pressure will lose a large amount of power, and will die over a short period of time; however, it will usually continue to produce some power during that time. Piston engines under the same circumstances are prone to seizing or breaking parts that almost certainly results in major internal damage of the engine and an instant loss of power. For this reason, Wankel engines are very well suited to snowmobiles, which often take users into remote places where a failure could result in frostbite or death, and aircraft, where abrupt failure is likely to lead to a crash or to force an improvised emergency ground landing or ditching (i.e. landing onto or into water by a plane not equipped with pontoons or other means for normal water landings).

Due to a 50% longer stroke duration compared to a four-cycle engine, there is more time to complete the combustion. This leads to greater suitability for direct injection. A Wankel rotary engine has stronger flows of air-fuel mixture and a longer operating cycle than a reciprocating engine, so it realises concomitantly thorough mixing of hydrogen and air. The result is a homogeneous mixture, which is crucial for hydrogen combustion.

Disadvantages

Although in two dimensions the seal system of a Wankel looks to be even simpler than that of a corresponding multi-cylinder piston engine, in three dimensions the opposite is true. As well as the rotor apex seals evident in the conceptual diagram, the rotor must also seal against the chamber ends.

Piston rings are not perfect seals: each has a gap to allow for expansion. The sealing at the Wankel apexes is less critical, as leakage is between adjacent chambers on adjacent strokes of the cycle, rather than to the crankcase. However, the less effective sealing of the Wankel is one factor reducing its efficiency, limiting its use mainly to applications such as racing engines and sports vehicles where neither efficiency nor long engine life are major considerations. Comparison tests have shown that the Mazda rotary powered RX-8 uses more fuel than a heavier vehicle powered by larger displacement V-8 engine for similar performance results.

The time available for fuel to be port-injected into a Wankel engine is significantly shorter, compared to four-stroke piston engines, due to the way the three chambers rotate. The fuel-air mixture cannot be pre-stored as there is no intake valve. Also the Wankel engine, compared to a piston engine, has 50% longer stroke duration. The four Otto cycles last 1080° for a Wankel engine versus 720° for a four-stroke reciprocating piston engine.

There are various methods of calculating the engine displacement of a Wankel. The Japanese regulations for calculating displacements for engine ratings use the volume displacement of one rotor face only, and the auto industry commonly accepts this method as the standard for calculating the displacement of a rotary. However, when compared on the basis of specific output, the convention results in large imbalances in favour of the Wankel motor.

For comparison purposes between a Wankel Rotary engine and a piston engine, displacement and corresponding power output can more accurately be compared on the basis of displacement per revolution of the eccentric shaft. A calculation of this form dictates that a two rotor Wankel displacing 654 cc per face will have a displacement of 1.3 litres per every rotation of the eccentric shaft (only two total faces, one face per rotor going through a full power stroke) and 2.6 litres after two revolutions (four total faces, two faces per rotor going through a full power stroke).

The results are directly comparable to a 2.6-litre piston engine with an even number of cylinders in a conventional firing order, which will likewise displace 1.3 litres through its power stroke after one revolution of the crankshaft, and 2.6 litres through its power strokes after two revolutions of the crankshaft. A Wankel Rotary engine is still a 4-stroke engine and pumping losses from non-power strokes still apply, but the absence of throttling valves and a 50% longer stroke duration result in a significantly lower pumping loss compared against a four-stroke reciprocating piston engine. Measuring a Wankel rotary engine in this way more accurately explains its specific output, as the volume of its air fuel mixture put through a complete power stroke per revolution is directly responsible for torque and thus power produced.

The trailing side of the rotary engine's combustion chamber develops a squeeze stream which pushes back the flamefront. With the conventional two-spark-plug or one-spark-plug system and homogenous mixture, this squeeze stream prevents the flame from propagating to the combustion chamber's trailing side in the mid and high engine speed ranges.

This is why there can be more carbon monoxide and unburnt hydrocarbons in a Wankel's exhaust stream. A side-port exhaust, as is used in the Renesis, avoids this because the unburned mixture cannot escape. The Mazda 26B avoided this issue through a 3-spark plug ignition system. (As a result, at the Le Mans 24 hour endurance race in 1991, the 26B had significantly lower fuel consumption than the competing reciprocating piston engines. All competitors had the same amount of fuel available, because of the Le Mans 24 hour limited fuel quantity rule.) A peripheral intake port gives the highest MEP, however, side intake porting produces a more steady idle.

All Mazda-made Wankel rotaries, including the new Renesis found in the RX-8, burn a small quantity of oil by design; it is metred into the combustion chamber to preserve the apex seals. Owners must periodically add small amounts of oil increasing running costs.

Applications—Automobile Racing

In the racing world, Mazda has had substantial success with two-rotor, three-rotor, and four-rotor cars. Private racers have also had considerable success with stock and modified Mazda Wankel-engine cars.

The Sigma MC74 powered by a Mazda 12A engine was the first engine and only team from outside Western Europe or the United States to finish the entire 24 hours of the 24 Hours of Le Mans race, in 1974. Mazda is the only team from outside Western Europe or the United States to have won Le Mans outright and the only non-piston engine ever to win Le Mans, which the company accomplished in 1991 with their four-rotor 787B (2,622 cc/160 cu in—actual displacement, rated by FIA formula at 4,708 cc/287 cu in). The following year, a planned rule change at Le Mans made the Mazda 787B ineligible to race due to weight advantages.

The Mazda RX-7 has won more IMSA races in its class than any other model of automobile, with its one hundredth victory on September 2, 1990. Following that, the RX-7 won its class in the IMSA 24 Hours of Daytona race ten years in a row, starting in 1982. The RX7 won the IMSA Grand Touring Under Two Liter (GTU) championship each year from 1980 through 1987, inclusive.

Formula Mazda Racing features open-wheel race cars with Mazda Wankel engines, adaptable to both oval tracks and road courses, on several levels of competition. Since 1991, the professionally organised Star Mazda Series has been the most popular format for sponsors, spectators, and upward bound drivers. The engines are all built by one engine builder, certified to produce the prescribed power, and sealed to discourage tampering. They are in a relatively mild state of racing tune, so that they are extremely reliable and can go years between motor rebuilds.

The Malibu Grand Prix chain, similar in concept to commercial recreational kart racing tracks, operates several venues in the United States where a customer can purchase several laps around a track in a vehicle very similar to open wheel racing vehicles, but powered by a small Curtiss-Wright rotary engine.

In engines having more than two rotors, or two rotor race engines intended for high-rpm use, a multi-piece eccentric shaft may be used, allowing additional bearings between rotors. While this approach does increase the complexity of the eccentric shaft design, it has been used successfully in the Mazda's production three-rotor 20B-REW engine, as well as many low volume production race engines. The C-111-2 4 Rotor Mercedes-Benz eccentric shaft for the KE Serie 70, Typ DB M950 KE409 is made in one piece. Mercedes-Benz used split bearings.

Motorcycle Engines

From 1974 to 1977 Hercules produced a limited number of motorcycles powered by Wankel engines.

In the early 1980s, David Garside's BSA twin-rotor engine reached production at Norton as the air-cooled twin-rotor Norton Classic. The Classic was succeeded by the later liquid-cooled Norton Commander and the Interpol2, a police version. The Norton engine later formed the basis for the MidWest AE series aero-engine. Norton used a Wankel engine in several models including the Norton F1, F1 Sports, RC588, RCW588, NRS588, most notably Steve Hislop riding to various victories on Norton's F1 in the TT in 1992. Norton has proposed a new 588 cc twin-rotor model called the NRV588 and a 700 cc version called the NRV700.

The Suzuki RE5 single-rotor motorcycle was produced from 1975 to 1976. It proved to be a complex design, with liquid cooling and oil cooling, and multiple lubrication and carburation systems. It worked well and was smooth, but being rather heavy and having a modest 62 bhp power output, it did not sell well.

Dutch motorcycle importer and manufacturer van Veen produced small quantities of their dual rotor Wankel-engined OCR-1000 between 1978 and 1980, using surplus Comotor engines.

In 1972, Yamaha introduced the 'RZ201' at the Tokyo Motor Show, a prototype with Wankel engine, weighing 220 kg and producing 60 hp from a 660 cc engine.

No motorcycles with Wankel engines have been produced for sale to the general public for road use since 1992.

Aircraft Engines

In principle, a Wankel engine should be ideal for light aircraft, as it is light, compact, almost vibrationless and has a high power-to-weight ratio. Further aviation benefits of a Wankel engine include: (i) rotors cannot seize, since rotor casings expand more than rotors; (ii) a Wankel is not susceptible to "shock-cooling" during descent; (iii) a Wankel does not require an enriched mixture for cooling at high power; (iv) having no reciprocating parts, it is less vulnerable to damage during "over-revving" (the main rev-limit being the strength of the main bearings). Unlike the case with cars (such as the NSU

Ro 80) and motorcycles, a Wankel aero-engine will (because of the time taken for pre-flight checks) inevitably be sufficiently warm before full power is asked of it. A Wankel aero-engine spends most of its working time at high power outputs, with little idling, so peripheral ports are ideal; and this makes feasible modular engines with more than two rotors. If "carb-icing" is an issue, plenty of waste engine heat is available to prevent trouble.

The first Wankel rotary-engine aircraft was the experimental Lockheed Q-Star civilian version of the United States Army's reconnaissance QT-2, basically a powered Schweizer sailplane, in 1968 or 1969. It was powered by a 185 hp (138 kW) Curtiss-Wright RC2-60 Wankel rotary engine; the same engine model was also flown in a Cessna Cardinal and other aeroplanes and a helicopter.

Aircraft Wankels have made something of a comeback in recent years. None of their advantages have been lost in comparison to other engines. They are increasingly being found in roles where their compact size and quiet operation is important, notably in drones, or UAVs. Many companies and hobbyists adapt Mazda rotary engines (taken from automobiles) to aircraft use; others, including Wankel GmbH itself, manufacture Wankel rotary engines dedicated for the purpose. One such use are the "Rotapower" engines in the Moller Skycar M400.

Wankel engines are also becoming increasingly popular in homebuilt experimental aircraft, such as the ARV Super2 which can be re-engined with the MidWest AE series aero-engine. Most are Mazda 12A and 13B automobile engines, converted to aviation use. This is a very cost-effective alternative to certified aircraft engines, providing engines ranging from 100 to 300 horsepower (220 kW) at a fraction of the cost of traditional engines. These conversions first took place in the early 1970s. With a number of these engines mounted on aircraft, as of 10 December 2006 the National Transportation Safety Board has only seven reports of incidents involving aircraft with Mazda engines, and none of these were a failure due to design or manufacturing flaws.

Peter Garrison, contributing editor for *Flying* magazine, has said that "the most promising engine for aviation use is the Mazda rotary." Mazdas have indeed worked well when converted for use in homebuilt aircraft. However, the real challenge in aviation is producing FAA-certified alternatives to the standard reciprocating engines that power most small general aviation aircraft. Mistral Engines, based in Switzerland, developed purpose-built rotaries for factory and retrofit installations on certified production aircraft. The G-190 and G-230-

TS rotary engines were already flying in the experimental market, and Mistral Engines hoped for FAA and JAA certification by 2011. As of June 2010, G-300 rotary engine development ceased, with the company citing a need for cash flow to complete development.

Mistral claims to have overcome the challenges of fuel consumption inherent in the rotary, at least to the extent that the engines are demonstrating specific fuel consumption within a few points of reciprocating engines of similar displacement. While fuel burn is still marginally higher than traditional engines, it is outweighed by other beneficial factors.

At the price of increased complication for a high pressure diesel type injection system, fuel consumption in the same range as small pre-chamber automotive and industrial diesels has been demonstrated with Curtiss-Wright's Stratified Charge multi-fuel engines, while preserving the aforementioned Wankel rotary advantages

Since Wankel engines operate at a relatively high rotational speed with relatively low torque, propeller aircraft must use a Propeller Speed Reduction Unit (PSRU) to keep their propellers within the proper speed range. Experimental aircraft with Wankel engines use PSRUs: for instance, the MidWest twin-rotor engine has a 2.95:1 reduction gearbox.

Pratt & Whitney Rocketdyne have been commissioned by DARPA to develop a diesel wankel engine for use in a prototype VTOL flying car called the "Transformer". The engine, based on an earlier UAV diesel wankel concept called 'EnduroCORE', will utilise wankel rotors of varying sizes on a shared eccentric shaft to increase efficiency. The engine is claimed to be a 'full-compression, full-expansion, diesel-cycle engine'. An October 28, 2010 patent from Pratt & Whitney Rocketdyne, describes a Wankel engine superficially similar to Rolls-Royce's earlier prototype that required an external air compressor to achieve high enough compression for diesel-cycle combustion. The design differs from Rolls-Royce's diesel wankel mainly by proposing an injector both in the exhaust passage between the combustor rotor and expansion rotor stages, and an injector in the expansion rotor's expansion chamber, for 'afterburning'.

The sailplane manufacturer Schleicher uses Wankel engines in their self-launching models.

Range Extender

Due to the compact size and low weight of a wankel engine, it has found its way into electric vehicles as range extenders. In 2010

Audi revealed that in their electric car, the A1 e-tron, they would have a small 250 cc Wankel engine running at 5,000 rpm that would recharge the car's batteries as needed.

In 2010 FEV Inc revealed that in their electric version of the Fiat 500 a wankel engine would be used as range extender as well.

The DA36 E-Star, an aircraft designed by Siemens, Diamond Aircraft and EADS, employs a series hybrid powertrain, with a wankel engine as internal combustion engine.

Other Uses

Small Wankel engines are being found increasingly in other roles, such as go-karts, personal water craft and auxiliary power units for aircraft. The Graupner/O.S. 49-PI is a 1.27 hp (947 W) 5 cc Wankel engine for model aeroplane use which has been in production essentially unchanged since 1970; even with a large muffler, the entire package weighs only 380 grams (13.4 ounces).

The simplicity of the Wankel makes it well-suited for mini, micro, and micro-mini engine designs. The Microelectromechanical systems (MEMS) Rotary Engine Lab at the University of California, Berkeley has been developing Wankel engines of down to 1 mm in diametre with displacements less than 0.1 cc. Materials include silicon and motive power includes compressed air. The goal is to eventually develop an internal combustion engine that will deliver 100 milliwatts of electrical power; the engine itself will serve as the rotor of the generator, with magnets built into the engine rotor itself.

The largest Wankel engine was built by Ingersoll-Rand; available in 550 hp (410 kW) one rotor and 1,100 hp (820 kW) two rotor versions, displacing 41 litres per rotor with a rotor approximately one metre in diametre. It was available between 1975 and 1985. It was derived from a previous, unsuccessful Curtiss-Wright design, which failed because of a well-known problem with all internal combustion engines: the fixed speed at which the flame front travels limits the distance combustion can travel from the point of ignition in a given time, and thereby limiting the maximum size of the cylinder or rotor chamber which can be used. This problem was solved by limiting the engine speed to only 1200 rpm and the use of natural gas as fuel; this was particularly well chosen, since one of the major uses of the engine was to drive compressors on natural gas pipelines. Yanmar Diesel of Japan, produced some small, charge-cooled rotor rotary engines for uses such as chainsaws and outboard engines, some of their contributions are that the LDR (rotor recess in the leading edge of combustion chamber) engines had better exhaust emissions profiles, and that reed-valve controlled intake ports improve part-load and low RPM performance.

Non-internal Combustion

In addition for use as an internal combustion engine, the basic Wankel design has also been utilised for gas compressors, and superchargers for internal combustion engines, but in these cases, although the design still offers advantages in reliability, the basic advantages of the Wankel in size and weight over the four-stroke internal combustion engine are irrelevant. In a design using a Wankel supercharger on a Wankel engine, the supercharger is twice the size of the engine.

The Wankel design is used in the seat belt pre-tensioner system of some Mercedes-Benz and Volkswagen cars. When the deceleration sensors sense a potential crash, small explosive cartridges are triggered electrically and the resulting pressurised gas feeds into tiny Wankel engines which rotate to take up the slack in the seat belt systems, anchoring the driver and passengers firmly in the seat before a collision.

Gas Turbine

A gas turbine, also called a combustion turbine, is a type of internal combustion engine. It has an upstream rotating compressor coupled to a downstream turbine, and a combustion chamber in-between. Energy is added to the gas stream in the combustor, where fuel is mixed with air and ignited. In the high pressure environment of the combustor, combustion of the fuel increases the temperature. The products of the combustion are forced into the turbine section. There, the high velocity and volume of the gas flow is directed through a nozzle over the turbine's blades, spinning the turbine which powers the compressor and, for some turbines, drives their mechanical output. The energy given up to the turbine comes from the reduction in the temperature and pressure of the exhaust gas.

Energy can be extracted in the form of shaft power, compressed air or thrust or any combination of these and used to power aircraft, trains, ships, generators, or even tanks.

History

- 150: Hero's Engine (*aeolipile*) — Apparently, Hero's steam engine was taken to be no more than a toy, and thus its full potential not realised for centuries.

- 1500: The “Chimney Jack” was drawn by Leonardo da Vinci: Hot air from a fire rises through a single-stage axial turbine rotor mounted in the exhaust duct of the fireplace and turning the roasting spit by gear/ chain connection.
- 1551: Taqi al-Din invented a rudimentary form of an impulse steam turbine, which he used to power a self-rotating spit.
- 1629: Jets of steam rotated an impulse turbine that then drove a working stamping mill by means of a bevel gear, developed by Giovanni Branca.
- 1678: Ferdinand Verbiest built a model carriage relying on a steam jet for power.
- 1791: A patent was given to John Barber, an Englishman, for the first true gas turbine. His invention had most of the elements present in the modern day gas turbines. The turbine was designed to power a horseless carriage.
- 1872: A gas turbine engine was designed by Dr. Franz Stolze, but the engine never ran under its own power.
- 1894: Sir Charles Parsons patented the idea of propelling a ship with a steam turbine, and built a demonstration vessel, the *Turbinia*, easily the fastest vessel afloat at the time. This principle of propulsion is still of some use.
- 1895: Three 4-ton 100 kW Parsons radial flow generators were installed in Cambridge Power Station, and used to power the first electric street lighting scheme in the city.
- 1899: Charles Gordon Curtis patented the first gas turbine engine in the USA (“Apparatus for generating mechanical power”, Patent No. US635,919).
- 1900: Sanford Alexander Moss submitted a thesis on gas turbines. In 1903, Dr. Moss became an engineer for General Electric’s Steam Turbine Department in Lynn, Massachusetts. While there, he applied some of his concepts in the development of the turbosuper charger. His design used a small turbine wheel, driven by exhaust gases, to turn a supercharger.
- 1903: A Norwegian, Ægidius Elling, was able to build the first gas turbine that was able to produce more power than needed to run its own components, which was considered an achievement in a time when knowledge about aerodynamics

was limited. Using rotary compressors and turbines it produced 11 hp (massive for those days). His work was later used by Sir Frank Whittle.

- 1906: The Armengaud-Lemale turbine engine in France with water-cooled combustion chamber.
- 1910: Holzwarth impulse turbine (pulse combustion) achieved 150 kilowatts.
- 1913: Nikola Tesla patents the Tesla turbine based on the boundary layer effect.
- 1918: One of the leading gas turbine manufacturers of today, General Electric, started their gas turbine division.
- 1920: The practical theory of gas flow through passages was developed into the more formal (and applicable to turbines) theory of gas flow past airfoils by Dr. A. A. Griffith.
- 1930: Sir Frank Whittle patented the design for a gas turbine for jet propulsion. The first successful use of his engine was in April 1937.
- 1932: BBC Brown, Boveri & Cie of Switzerland starts selling axial compressor and turbine turbosets as part of the turbocharged steam generating Velox boiler. Following the gas turbine principle, the steam evapouration tubes are arranged within the gas turbine combustion chamber; the first Velox plant was erected in Mondeville, France.
- 1934: Raúl Pateras de Pescara patented the free-piston engine as a gas generator for gas turbines.
- 1936: Hans von Ohain and Max Hahn in Germany were developing their own patented engine design.
- 1939: First 4 MW utility power generation gas turbine from BBC Brown, Boveri & Cie. for an emergency power station in Neuchâtel, Switzerland.

Theory of Operation

Gases passing through an ideal gas turbine undergo three thermodynamic processes. These are isentropic compression, isobaric (constant pressure) combustion and isentropic expansion. Together these make up the Brayton cycle.

In a practical gas turbine, gases are first accelerated in either a centrifugal or radial compressor. These gases are then slowed using a diverging nozzle known as a diffuser; these processes increase the pressure and temperature of the flow. In an ideal system this is isentropic.

However, in practice energy is lost to heat, due to friction and turbulence. Gases then pass from the diffuser to a combustion chamber, or similar device, where heat is added. In an ideal system this occurs at constant pressure (isobaric heat addition).

As there is no change in pressure the specific volume of the gases increases. In practical situations this process is usually accompanied by a slight loss in pressure, due to friction.

Finally, this larger volume of gases is expanded and accelerated by nozzle guide vanes before energy is extracted by a turbine. In an ideal system these are gases expanded isentropically and leave the turbine at their original pressure.

In practice this process is not isentropic as energy is once again lost to friction and turbulence. If the device has been designed to power a shaft as with an industrial generator or a turboprop, the exit pressure will be as close to the entry pressure as possible.

In practice it is necessary that some pressure remains at the outlet in order to fully expel the exhaust gases. In the case of a jet engine only enough pressure and energy is extracted from the flow to drive the compressor and other components. The remaining high pressure gases are accelerated to provide a jet that can, for example, be used to propel an aircraft.

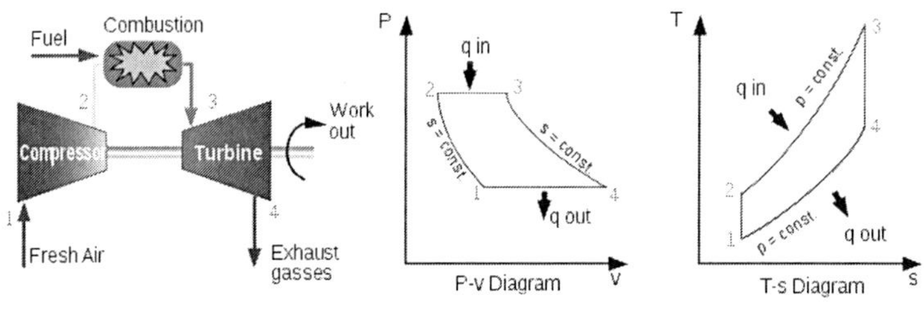

Figure 2: Brayton cycle

As with all cyclic heat engines, higher combustion temperatures can allow for greater efficiencies. However, temperatures are limited by ability of the steel, nickel, ceramic, or other materials that make up the engine to withstand high temperatures and stresses. To combat this many turbines feature complex blade cooling systems.

As a general rule, the smaller the engine the higher the rotation rate of the shaft(s) must be to maintain tip speed. Blade tip speed determines the maximum pressure ratios that can be obtained by the turbine and the compressor. This in turn limits the maximum power and efficiency that can be obtained by the engine.

In order for tip speed to remain constant, if the diametre of a rotor is reduced by half, the rotational speed must double. For example large Jet engines operate around 10,000 rpm, while micro turbines spin as fast as 500,000 rpm. Mechanically, gas turbines can be considerably less complex than internal combustion piston engines. Simple turbines might have one moving part: the shaft/compressor/turbine/alternative-rotor assembly, not counting the fuel system.

However, the required precision manufacturing for components and temperature resistant alloys necessary for high efficiency often make the construction of a simple turbine more complicated than piston engines.

More sophisticated turbines (such as those found in modern jet engines) may have multiple shafts (spools), hundreds of turbine blades, movable stator blades, and a vast system of complex piping, combustors and heat exchangers. Thrust bearings and journal bearings are a critical part of design. Traditionally, they have been hydrodynamic oil bearings, or oil-cooled ball bearings. These bearings are being surpassed by foil bearings, which have been successfully used in micro turbines and auxiliary power units.

Types of Gas Turbines

Jet Engines

Airbreathing jet engines are gas turbines optimised to produce thrust from the exhaust gases, or from ducted fans connected to the gas turbines.

Jet engines that produce thrust primarily from the direct impulse of exhaust gases are often called turbojets, whereas those that generate most of their thrust from the action of a ducted fan are often called turbofans or (rarely) fan-jets. Gas turbines are also used in many liquid propellant rockets, the gas turbines are used to power a turbopump to permit the use of lightweight, low pressure tanks, which saves considerable dry mass.

Turboprop Engines

A turboprop engine is a type of turbine engine which drives an external aircraft propeller using a reduction gear. Turboprop engines are generally used on small subsonic aircraft, but some large military and civil aircraft, such as the Airbus A400M, Lockheed L-188 Electra and Tupolev Tu-95, have also used turboprop power.

Aeroderivative Gas Turbines

Aeroderivatives are also used in electrical power generation due to their ability to be shut down, and handle load changes more quickly than industrial machines.

They are also used in the marine industry to reduce weight. The General Electric LM2500, General Electric LM6000, Rolls-Royce RB211 and Rolls-Royce Avon are common models of this type of machine.

Amateur Gas Turbines

Increasing numbers of gas turbines are being used or even constructed by amateurs.

In its most straightforward form, these are commercial turbines acquired through military surplus or scrapyard sales, then operated for display as part of the hobby of engine collecting. In its most extreme form, amateurs have even rebuilt engines beyond professional repair and then used them to compete for the Land Speed Record.

The simplest form of self-constructed gas turbine employs an automotive turbocharger as the core component. A combustion chamber is fabricated and plumbed between the compressor and turbine sections.

More sophisticated turbojets are also built, where their thrust and light weight are sufficient to power large model aircraft. The Schreckling design constructs the entire engine from raw materials, including the fabrication of a centrifugal compressor wheel from plywood, epoxy and wrapped carbon fibre strands.

Several small companies now manufacture small turbines and parts for the amateur. Most turbojet-powered model aircraft are now using these commercial and semi-commercial microturbines, rather than a Schreckling-like home-build.

Auxiliary Power Units

APUs are small gas turbines designed for auxiliary power of larger machines, such as those inside an aircraft. They supply compressed air for aircraft ventilation (with an appropriate compressor design), start-up power for larger jet engines, and electrical and hydraulic power.

Industrial Gas Turbines for Power Generation

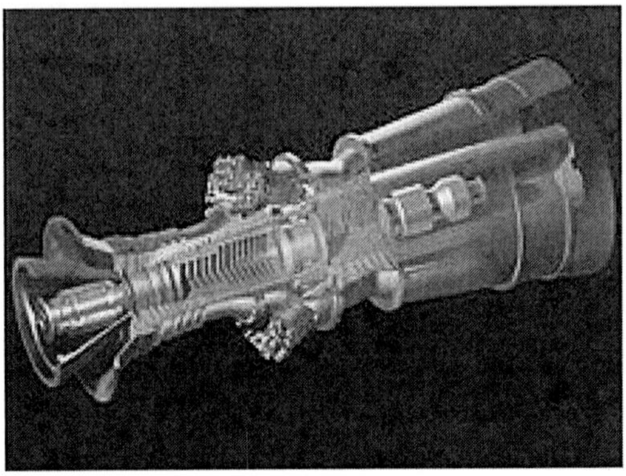

Figure 3: GE H series power generation gas turbine: in combined cycle configuration, this 480-megawatt unit has a rated thermal efficiency of 60%.

Industrial gas turbines differ from aeroderivative in that the frames, bearings, and blading are of heavier construction. Industrial gas turbines range in size from truck-mounted mobile plants to enormous, complex systems.

They can be particularly efficient—up to 60%—when waste heat from the gas turbine is recovered by a heat recovery steam generator to power a conventional steam turbine in a combined cycle configuration. They can also be run in a cogeneration configuration: the exhaust is used for space or water heating, or drives an absorption chiller for cooling or refrigeration. Such engines require a dedicated enclosure, both to protect the engine from the elements and the operators from the noise.

The construction process for gas turbines can take as little as several weeks to a few months, compared to years for base load power plants. Their other main advantage is the ability to be turned on and off within minutes, supplying power during peak demand.

Since single cycle (gas turbine only) power plants are less efficient than combined cycle plants, they are usually used as peaking power plants, which operate anywhere from several hours per day to a few dozen hours per year, depending on the electricity demand and the generating capacity of the region.

In areas with a shortage of base load and load following power plant capacity or low fuel costs, a gas turbine power plant may regularly operate during most hours of the day. A large single cycle gas turbine typically produces 100 to 400 megawatts of power and have 35–40% thermal efficiency.

Compressed Air Energy Storage

Compressed Air Energy Storage (CAES) is a way to store energy generated at one time for use at another time. At utility scale, energy generated during periods of low energy demand (off-peak) can be released to meet higher demand (peak load) periods.

Types

Compression of air generates a lot of heat. The air is warmer after compression. Decompression requires heat. If no extra heat is added, the air will be much colder after decompression. If the heat generated during compression can be stored and used again during decompression, the efficiency of the storage improves considerably. There are three ways in which a CAES system can deal with the heat. Air storage can be adiabatic, diabatic, or isothermic:

- Adiabatic storage retains the heat produced by compression and returns it to the air when the air is expanded to generate power. This is a subject of ongoing study, with no utility scale plants as of 2010. Its theoretical efficiency approaches 100% for large and/or rapidly cycled devices and/or perfect thermal insulation, but in practice round trip efficiency is expected to be 70%. Heat can be stored in a solid such as concrete or stone, or more likely in a fluid such as hot oil (up to 300 °C) or molten salt solutions (600 °C).
- Diabatic storage dissipates the extra heat with intercoolers (thus approaching isothermal compression) into the atmosphere as waste. Upon removal from storage, the air must be re-heated prior to expansion in the turbine to power a generator which can be accomplished with a natural gas fired burner for utility grade storage or with a heated metal mass. The lost heat degrades efficiency, but this approach is simpler and is thus far the only system which has been implemented commercially. The McIntosh, Alabama CAES plant requires 2.5 MJ of electricity and 1.2 MJ lower heating value (LHV) of gas for each megajoule of energy output. A General Electric 7FA 2x1 combined cycle plant, one of the most efficient natural gas plants in operation, uses 6.6 MJ (LHV) of gas per kW–h generated, a 54% thermal efficiency comparable to the McIntosh 6.8 MJ, at 53% thermal efficiency.
- Isothermal compression and expansion approaches attempt to maintain operating temperature by constant heat exchange to

the environment. They are only practical for low power levels, without very effective heat exchangers. The theoretical efficiency of isothermal energy storage approaches 100% for small and/or slowly cycled devices and/or perfect heat transfer to the environment. In practice neither of these perfect thermodynamic cycles are obtainable, as some heat losses are unavoidable.

A different, highly efficient arrangement, which fits neatly into none of the above categories, uses high, medium and low pressure pistons in series, with each stage followed by an airblast venturi pump that draws ambient air over an air-to-air (or air-to-seawater) heat exchanger between each expansion stage. Early compressed air torpedo designs used a similar approach, substituting seawater for air. The venturi warms the exhaust of the preceding stage and admits this preheated air to the following stage. This approach was widely adopted in various compressed air vehicles such as H. K. Porter, Inc's mining locomotives and trams. Here the heat of compression is effectively stored in the atmosphere (or sea) and returned later on.

Compression can be done with electrically powered turbo-compressors and expansion with turbo 'expanders' or air engines driving electrical generators to produce electricity.

The storage vessel is often an underground cavern created by solution mining (salt is dissolved in water for extraction) or by utilising an abandoned mine. Plants operate on a daily cycle, charging at night and discharging during the day. Compressed air energy storage can also be employed on a smaller scale such as exploited by air cars and air-driven locomotives, and also by the use of high-strength carbon-fibre air storage tanks.

History

City-wide compressed air energy systems have been built since 1870. Cities such as Paris, Birmingham, England, Rixdorf, Offenbach, Dresden in Germany and Buenos Aires, Argentina installed such systems. Victor Popp constructed the first systems to power clocks by sending a pulse of air every minute to change the pointer. They quickly evolved to deliver power to homes and industry. As of 1896, the Paris system had 2.2 MW of generation distributed at 550 kPa in 50 km of air pipes for motors in light and heavy industry. Usage was measured by metres. The systems were the main source of house-delivered energy in these days and also powered the machines of dentists, seamstresses, printing facilities and bakeries. The first utility-scale compressed air energy storage project was the 290 megawatt Huntorf plant in Germany (1978). The second was the 110 megawatt McIntosh plant in Alabama (1991).

Both of these projects use salt domes for air caverns. Currently under development is the Iowa Stored Energy Park, a 270 megawatt project which will use aquifer-based air storage. A 300 MW project utilising depleted gas storage is being developed in California, and a 150 MW salt-based project is under development in upstate New York. The first adiabatic CAES project, a 200 megawatt facility called ADELE, is planned for construction in Germany in 2013.

Isothermal Storage Physics

The isothermal process maintains constant temperature. The heat that compression generates must flow to the environment for the temperature to remain constant. In practice this is often not the case, because proper intercooling requires a compact internal heat exchanger that is optimised for high heat transfer and low pressure drop.

Without an internal heat exchanger, isothermal compression can be approximated only at low flow rates, particularly for small systems. Small compressors have higher inherent heat exchange, due to a higher ratio of surface area to volume. Nevertheless it is useful to describe the limiting case of ideal isothermal compression of an ideal gas:

The ideal gas law for an isothermal process is:

$$PV = nRT = \text{constant}$$

By the definition of work, where A and B are the initial and final states of the system:

$$W_{A\to B} = \int_{V_A}^{V_B} P\,dV = \int_{V_A}^{V_B} \frac{nRT}{V}\,dV = nRT \int_{V_A}^{V_B} \frac{1}{V}\,dV$$

$$= nRT(\ln V_B - \ln V_A) = nRT \ln \frac{V_B}{V_A} = nRT \ln \frac{P_A}{P_B} = PV \ln \frac{P_A}{P_B}$$

where, $P_A V_A = P_B V_B$, and so, $\frac{V_B}{V_A} = \frac{P_A}{P_B}$

P is the absolute pressure,

V is the volume of the vessel,

n is the amount of substance of gas,

R is the ideal gas constant,

T is the absolute temperature,

W is the energy stored or released.

This amounts to about 2.271 $\ln(P_A/P_B)$ kJ at 0 °C (273 K) or 2.478 $\ln(P_A/P_B)$ kilojoules (kJ) at 25 °C (298 K), per mole, or simply 100 $\ln(P_A/P_B)$ kJ/m^3 of gas (at 0.1 megapascal = approx. atmospheric pressure).

Isothermal processes are thermodynamically reversible. To the extent the compression process is isothermal, its efficiency approaches 100%. The above equation represents the maximum energy that can be stored. To be more accurate, that energy isn't stored into compressed air. Instead the energy given to compress the air is converted into heat that is expelled to the ambient. So the energy is stored into the ambient while the air is just the vehicle to do that.

This is true because according to thermodynamics (for the case of a perfect gas) the internal energy of the gas is given by E=(df/2)*N*k*T where df are the degrees of freedom, N the number of particles, k the Boltzmann constant. So a isothermal compression just diminishes the entropy of air while converting work to heat that is stored in the ambient. Inversely that air may be used at a later time to convert heat into work. But the heat must be taken from the ambient or somewhere else. In practice, no process is perfectly isothermal and the compressors and motors will have heat-related energy losses.

Under adiabatic compression, some of the compression work goes into heating the gas. If this heat is then lost to the surroundings, and the same quantity of heat is not added to the gas upon expansion, efficiency is reduced. Energy storage systems often use large underground caverns. This is the preferred system design, due to the very large volume, and thus the large quantity of energy that can be stored with only a small pressure change. The cavern space can be compressed adiabatically with little temperature change and heat loss.

Practical Constraints in Transportation

Energy Density and Efficiency

Compressing air heats it and expanding it cools it. Therefore practical air engines require heat exchangers in order to avoid excessively high or low temperatures and even so don't reach ideal constant temperature conditions. Nevertheless it is useful to describe the maximum energy storable using the isothermal case, which works out to about 100 kJ/m^3 [ln(P_A/P_B)]. Thus if 1.0 m^3 of ambient air is very slowly compressed into a 5 L bottle at 20 MPa, the potential energy stored is 530 kJ. A highly efficient air motor can transfer this into kinetic energy if it runs very slowly and manages to expand the air from its initial 20 MPa pressure down to 100 kPa (bottle completely "empty" at ambient pressure). Achieving high efficiency is a technical challenge both due to nonlinear energy storage and thermodynamic considerations.

If the bottle above is emptied to 1 MPa, the extractable energy is about 300 kJ at the motor shaft.

A standard 20 MPa, 5 L steel bottle has a mass of 7.5 kg, a superior one 5 kg. High-tensile strength fibres such as carbon-fibre or Kevlar can weigh below 2 kg in this size, consistent with the legal safety codes. One cubic metre of air at 20°C has a mass of 1.225 kg. Thus, theoretical energy densities are from roughly 70 kJ/kg at the motor shaft for a plain steel bottle to 180 kJ/kg for an advanced fibre-wound one, whereas practical achievable energy densities for the same containers would be from 40 to 100 kJ/kg.

Comparison with Batteries

Advanced fibre-reinforced bottles are comparable to the rechargeable lead-acid battery in terms of energy density. Advanced battery systems are several times better. Batteries also provide nearly constant voltage over their entire charge level, whereas the pressure varies greatly while using a pressure vessel from full to empty. It is technically challenging to design air engines to maintain high efficiency and sufficient power over a wide range of pressures. Compressed air can transfer power at very high flux rates, which meets the principal acceleration and deceleration objectives of transportation systems, particularly for hybrid vehicles.

Compressed air systems have advantages over conventional batteries including longer lifetimes of pressure vessels and lower material toxicity. Newer battery designs such as those based on Lithium Iron Phosphate chemistry suffer from neither of these problems. Compressed air costs are potentially lower; however advanced pressure vessels are costly to develop and safety-test and at present are more expensive than mass-produced batteries.

As with electric storage technology, compressed air is only as "clean" as the source of the energy that it stores. Life cycle assessment addresses the question of overall emissions from a given energy storage technology combined with a given mix of generation on a power grid.

Safety

As with most technologies, compressed air has safety concerns, mainly catastrophic tank rupture. Safety codes make this a rare occurrence at the cost of higher weight. Codes may limit the legal working pressure to less than 40% of the rupture pressure for steel bottles (safety factor of 2.5), and less than 20% for fibre-wound bottles (safety factor of 5). Commercial designs adopt the ISO 11439 standard. High pressure bottles are fairly strong so that they generally do not rupture in vehicle crashes.

Compressed-air Vehicle

A compressed-air vehicle (CAV) is powered by an air engine, using compressed air, which is stored in a tank. Instead of mixing fuel with air and burning it in the engine to drive pistons with hot expanding gases, compressed-air vehicles use the expansion of compressed air to drive their pistons. One manufacturer claims to have designed an engine that is 90 percent efficient. Compressed-air propulsion may also be incorporated in hybrid systems, e.g., battery electric propulsion and fuel tanks to recharge the batteries. This kind of system is called a hybrid-pneumatic electric propulsion. Additionally, regenerative braking can also be used in conjunction with this system.

Technology

Engine

One can buy the vehicle with the engine or buy an engine to be installed in the vehicle. Typical air engines use one or more expander pistons or rotary expander like the Quasiturbine. In some applications it is advantageous to heat the air, or the engine, to increase the range or power.

Tanks

The tanks must be designed to safety standards appropriate for a pressure vessel, such as ISO 11439.

The storage tank may be made of:

- steel,
- aluminium,
- carbon fibre,
- Kevlar,
- other materials, or combinations of the above.

The fibre materials are considerably lighter than metals but generally more expensive. Metal tanks can withstand a large number of pressure cycles, but must be checked for corrosion periodically.

One company stores air in tanks at 4,500 pounds per square inch (about 30 MPa) and hold nearly 3,200 cubic feet (around 90 cubic metres) of air.

The tanks may be refilled at a service station equipped with heat exchangers, or in a few hours at home or in parking lots, plugging the car into the electrical grid via an on-board compressor. The cost of driving such a car is typically projected to be around €0.75 per 100 km, with a complete refill at the "tank-station" at about US$3.

Compressed Air

Compressed air has a low energy density. In 300 bar containers, about 0.1 MJ/L and 0.1 MJ/kg is achievable, comparable to the values of electrochemical lead-acid batteries. While batteries can somewhat maintain their voltage throughout their discharge and chemical fuel tanks provide the same power densities from the first to the last litre, the pressure of compressed air tanks falls as air is drawn off. A consumer-automobile of conventional size and shape typically consumes 0.3-0.5 kWh (1.1-1.8 MJ) at the drive shaft per mile of use, though unconventional sizes may perform with significantly less.

Emission Output

Like other non-combustion energy storage technologies, an air vehicle displaces the emission source from the vehicle's tail pipe to the central electrical generating plant. Where emissions-free sources are available, net production of pollutants can be reduced. Emission control measures at a central generating plant may be more effective and less costly than treating the emissions of widely dispersed vehicles.

Since the compressed air is filtered to protect the compressor machinery, the air discharged has less suspended dust in it, though there may be carry-over of lubricants used in the engine.

History

Compressed air has been used since the 19th century to power mine locomotives and trams in cities such as Paris (via a central, city-level, compressed air energy distribution system), and was previously the basis of naval torpedo propulsion.

During the construction of the Gotthardbahn from 1872 to 1882, pneumatic locomotives were used in the construction of the Gotthard Rail Tunnel and other tunnels of the Gotthardbahn.

In 1903, the Liquid Air Company located in London England manufactured a number of compressed-air and liquified-air cars. The major problem with these cars and all compressed-air cars is the lack of torque produced by the "engines" and the cost of compressing the air.

Recently several companies have started to develop compressed air cars, although none have been released to the public, or have been tested by third parties.

Advantages

Compressed-air vehicles are comparable in many ways to electric vehicles, but use compressed air to store the energy instead of batteries. Their potential advantages over other vehicles include:

- Much like electrical vehicles, air powered vehicles would ultimately be powered through the electrical grid. Which makes it easier to focus on reducing pollution from one source, as opposed to the millions of vehicles on the road.
- Transportation of the fuel would not be required due to drawing power off the electrical grid. This presents significant cost benefits. Pollution created during fuel transportation would be eliminated.
- Compressed-air technology reduces the cost of vehicle production by about 20%, because there is no need to build a cooling system, fuel tank, Ignition Systems or silencers.
- Air, on its own, is non-flammable.
- The engine can be massively reduced in size.
- The engine runs on cold or warm air, so can be made of lower strength light weight material such as aluminium, plastic, low friction teflon or a combination.
- Low manufacture and maintenance costs as well as easy maintenance.
- Compressed-air tanks can be disposed of or recycled with less pollution than batteries.
- Compressed-air vehicles are unconstrained by the degradation problems associated with current battery systems.
- The air tank may be refilled more often and in less time than batteries can be recharged, with re-filling rates comparable to liquid fuels.

- Lighter vehicles cause less damage to roads, resulting in lower maintenance cost.
- The price of filling air powered vehicles is significantly cheaper than petrol, diesel or biofuel. If electricity is cheap, then compressing air will also be relatively cheap.

Disadvantages

Like the modern car and most household appliances, the principal disadvantage is the indirect use of energy. Energy is used to compress air, which - in turn - provides the energy to run the motor. Any conversion of energy between forms results in loss.

For conventional combustion motor cars, the energy is lost when oil is converted to usable fuel - including drilling, refinement, labour, storage, eventually transportation to the end-user. For compressed-air cars, energy is lost when electrical energy is converted to compressed air.

- When air expands, as it would in the engine, it cools dramatically (Charles's law) and must be heated to ambient temperature using a heat exchanger similar to the Intercooler used for internal combustion engines. The heating is necessary in order to obtain a significant fraction of the theoretical energy output. The heat exchanger can be problematic. While it performs a similar task to the Intercooler, the temperature difference between the incoming air and the working gas is smaller. In heating the stored air, the device gets very cold and may ice up in cool, moist climates.
- Refuelling the compressed-air container using a home or low-end conventional air compressor may take as long as 4 hours though the specialised equipment at service stations may fill the tanks in only 3 minutes.
- Tanks get very hot when filled rapidly. SCUBA tanks are sometimes immersed in water to cool them down when they are being filled. That would not be possible with tanks in a car and thus it would either take a long time to fill the tanks, or they would have to take less than a full charge, since heat drives up the pressure.
- Early tests have demonstrated the limited storage capacity of the tanks; the only published test of a vehicle running on compressed air alone was limited to a range of 7.22 km (4 mi).

- A 2005 study demonstrated that cars running on lithium-ion batteries out-perform both compressed-air and fuel cell vehicles more than threefold at same speeds. MDI has recently claimed that an air car will be able to travel 140 km (87 mi) in urban driving, and have a range of 80 km (50 mi) with a top speed of 110 km/h (68 mph) on highways, when operating on compressed air alone.

Possible Improvements

- Compressed-air vehicles operate to a thermodynamic process as air cools down when expanding and heats up when being compressed. As it is not possible in practice to use a theoretically ideal process, losses occur and improvements may involve reducing these, e.g., by using large heat exchangers in order to use heat from the ambient air and at the same time provide air cooling in the passenger compartment. At the other end, the heat produced during compression can be stored in water systems, physical or chemical systems and reused later.
- It may be possible to store compressed air at lower pressure using an absorption material within the tank. Absorption materials such as Activated carbon, or a metal organic framework is used to store compressed natural gas at 500 psi instead of 4500 psi, which amounts to a large energy saving.

Vehicles

Trikes

Three mechanical engineering students from San Jose State University; Daniel Mekis, Dennis Schaaf and Andrew Merovich, designed and built a bike that runs on compressed air. The total cost of the prototype was under $1000 and was sponsored by Sunshops and NO DIG NO RIDE. The top speed of the maiden voyage in May 2009 was 23 mph.

Three Wheeler

"Ku:Rin" named air-compressed three wheeler vehicle is created by TOYOTA in 2011. The speciality about this vehicle is it has registered a record breaking highest speed 129.2 km/h (80mph) even if it has engine which uses only compressed-air. This car was developed by the toyota companies "Dream car workshop". This car is nicknamed as "sleek rocket" , or "pencil shaped rocket".

Motorcycles

A compressed-air powered motorcycle, called the Green Speed Air Powered Motorcycle was made by Edwin Yi Yuan, based on the Suzuki GP100 and using the Angelo Di Pietro compressed air engine.

Mopeds

As part of the TV-show Planet Mechanics, Jem Stansfield and Dick Strawbridge have converted a regular scooter to a compressed air moped. This has been done by equipping the scooter with a compressed-air engine and air tank.

Cars

Several companies are investigating and producing prototypes, and one plans to offer cars late 2009, early 2010.

Buses

MDI makes MultiCATs vehicle that can be used as buses or trucks. RATP has also already expressed an interest in the compressed-air pollution-free bus.

Locomotives

Compressed-air locomotives are a kind of fireless locomotive and have been used in mining and tunnel boring.

Trams

Various compressed-air-powered trams were trialled, starting in 1876.

Watercraft and Aircraft

Currently, no water or air vehicles exist that make use of the air engine. Historically certain torpedoes were propelled by compressed-air engines.

History

Air engines have been used since the 19th century to power mine locomotives, pumps, drills and trams, via centralised, city-level, distribution.

Racecars use compressed air to start its internal combustion engine (ICE).

Many people have been working on the idea of compressed air vehicles with renewed interest since the 1990 oil price shock.

Pneumatic Motor

A pneumatic motor or compressed air engine is a type of motor which does mechanical work by expanding compressed air. Pneumatic motors generally convert the compressed air to mechanical work through either linear or rotary motion. Linear motion can come from either a diaphragm or piston actuator, while rotary motion is supplied by either a vane type air motor or piston air motor.

Pneumatic motors have existed in many forms over the past two centuries, ranging in size from hand held turbines to engines of up to several hundred horsepower.

Some types rely on pistons and cylinders, others use turbines. Many compressed air engines improve their performance by heating the incoming air, or the engine itself. Pneumatic motors have found widespread success in the hand-held tool industry and continual attempts are being made to expand their use to the transportation industry. However, pneumatic motors must overcome inefficiencies before being seen as a viable option in the transportation industry.

Classification

Linear

In order to achieve linear motion from compressed air, a system of pistons is most commonly used. The compressed air is fed into an air-tight chamber that houses the shaft of the piston. Also inside this chamber a spring is coiled around the shaft of the piston in order to hold the chamber completely open when air is not being pumped into the chamber.

As air is fed into the chamber the force on the piston shaft begins to overcome the force being exerted on the spring. As more air is fed into the chamber, the pressure increases and the piston begins to move down the chamber. When it reaches its maximum length the air pressure is released from the chamber and the spring completes the cycle by closing off the chamber to return to its original position.

Piston motors are the most commonly used in hydraulic systems. Essentially, piston motors are the same as hydraulic pumps except they are used to convert hydraulic energy into mechanical energy. Piston motors are often used in series of two, three, four, five, or six cylinders that are enclosed in a housing. This allows for more power to be delivered by the pistons because several motors are in sync with each other at certain times of their cycle.

Rotary

Another type of pneumatic motor, known as a rotary vane motor, uses air to produce rotational motion to a shaft. The rotating element is a slotted rotor which is mounted on a drive shaft. Each slot of the rotor is fitted with a freely sliding rectangular vane. The vanes are extended to the housing walls using springs, cam action, or air pressure, depending on the motor design. Air is pumped through the motor input which pushes on the vanes creating the rotational motion of the central shaft. Rotation speeds can vary between 100 and 25,000 rpm depending on several factors which including the amount of air pressure at the motor inlet and the diametre of the housing.

Rotary motion vane-type air motors are used to start large industrial diesel or natural gas engines. Stored energy in the form of compressed air, nitrogen or natural gas enters the sealed motor chamber and exerts pressure against the vanes of a rotor. Much like a windmill, this causes the rotor to turn at high speed.

Because the engine flywheel requires a great deal of torque to start the engine, reduction gears are used. Reduction gears create high torque levels with the lower amounts of energy input. These reduction gears allow for sufficient torque to be generated by the engine flywheel while it is engaged by the pinion gear of the air motor or air starter.

Application

A widespread application of small pneumatic motors is in hand-held tools, power ratchet wrenches, drills, sanders, grinders, cutters, and so on. Though overall energy efficiency of pneumatics tools is low and they require access to a compressed-air source, there are several advantages over electric tools. They offer greater power density (a smaller pneumatic motor can provide the same amount of power as a larger electric motor), do not require an axillary speed controller (adding to its compactness), generate less heat, and can be used in more volatile atmospheres as they do not require electric power.

Historically, many individuals have tried to apply pneumatic motors to the transportation industry. Guy Negre, CEO and founder of Zero Pollution Motors, has pioneered this field since the late 1980s. Recently Engineair has also developed a rotary motor for use in automobiles. Engineair places the motor immediately beside the wheel of the vehicle and uses no intermediate parts to transmit motion which means almost all of the motor's energy is used to rotate the wheel.

History in Transportation

The pneumatic motor was first applied to the field of transportation in the mid-19th century. Though little is known about the first recorded compressed-air vehicle, it is said that the Frenchmen Andraud and Tessie of Motay ran a car powered by a pneumatic motor on a test track in Chaillot, France, on July 9, 1840. Although the car test was reported to have been successful, the pair didn't explore further expansion of the design.

The first successful application of the pneumatic motor in transportation was the Mekarski system air engine used in locomotives. Mekarski's innovative engine overcame cooling that accompanies air compression by heating air in a small boiler prior to use. The Tramway de Nantes, located in Nantes, France, was noted for being the first to use Mekarski engines to power their fleet of locomotives. The tramway began operation on December 13, 1879, and continues to operate today, although the pneumatic trams were replaced in 1917 by more efficient and modern electrical trams.

American Charles Hodges also found success with pneumatic motors in the locomotive industry. In 1911 he designed a pneumatic locomotive and sold the patent to the H. K. Porter Company in Pittsburgh for use in coal mines. Because pneumatic motors do not use combustion they were a much safer option in the coal industry.

Many companies claim to be developing Compressed air cars, but none are actually available for purchase or even independent testing.

Tools

Impact wrenches, drills, firearms, die grinders, dental drills and other pneumatic tools use a variety of air engines or motors. These include vane type pumps, turbines and pistons.

Torpedoes

Most successful early forms of self propelled torpedoes used high pressure compressed air, although this was superseded by internal or external combustion engines, steam engines, or electric motors.

Railways

Compressed air engines were used in trams and shunters, and eventually found a successful niche in mining locomotives, although eventually they were replaced by electric trains, underground. Over the years designs increased in complexity, resulting in a triple expansion engine with air-to-air reheaters between each stage.

Aircraft

Transport category aeroplanes, such as commercial airliners, use compressed air starters to start the main engines. The air is supplied by the load compressor of the aircraft's auxiliary power unit, or by ground equipment.

Automotive

There is currently some interest in developing air cars. Several engines have been proposed for these, although none have demonstrated the performance and long life needed for personal transport.

Energine

The Energine Corporation was a South Korean company that claimed to deliver fully assembled cars running on a hybrid compressed air and electric engine. The compressed-air engine is used to activate an alternator, which extends the autonomous operating capacity of the car. The CEO was arrested for fraudulently promoting air motors with false claims

EngineAir

EngineAir, an Australian company, is making a rotary engine powered by compressed air, called The Di Pietro motor. The Di Pietro motor concept is based on a rotary piston. Different from existing rotary engines, the Di Pietro motor uses a simple cylindrical rotary piston (shaft driver) which rolls, with little friction, inside the cylindrical stator.

It can be used in boat, cars, carriers and other vehicles. Only 1 psi (H" 6,8 kPa) of pressure is needed to overcome the friction. The engine was also featured on the ABC's New Inventors program in Australia on 24 March 2004.

K'Airmobiles

K'Airmobiles vehicles were intended to be commercialised from a project developed in France in 2006-2007 by a small group of researchers. However, the project has not been able to gather the necessary funds.

People should note that, meantime, the team has recognised the physical impossibility to use on-board stored compressed air due to its poor energy capacity and the thermal losses resulting from the expansion of the gas.

These days, using the patent pending 'K'Air Generator', converted to work as a compressed-gas motor, the project should be launched in 2010, thanks to a North American group of investors, but for the purpose of developing first a green energy power system.

MDI

In the original Nègre air engine, one piston compresses air from the atmosphere to mix with the stored compressed air (which will cool drastically as it expands). This mixture drives the second piston, providing the actual engine power. MDI's engine works with constant torque, and the only way to change the torque to the wheels is to use a pulley transmission of constant variation, losing some efficiency. When vehicle is stopped, MDI's engine had to be on and working, losing energy. In 2001-2004 MDI switched to a design similar to that described in Regusci's patents, which date back to 1990.

It has been reported in 2008 that Indian car manufacturer Tata was looking at an MDI compressed air engine as an option on its low priced Nano automobiles. Tata announced in 2009 that the compressed air car was proving difficult to develop due to its low range and problems with low engine temperatures.

Quasiturbine

The Pneumatic Quasiturbine engine is a compressed air pistonless rotary engine using a rhomboidal-shaped rotor whose sides are hinged at the vertices.

The Quasiturbine has demonstrated as a pneumatic engine using stored compressed air. It can also take advantage of the energy amplification possible from using available external heat, such as solar energy.

The Quasiturbine rotates from pressure as low as 0.1 atm (1.47psi).

Since the Quasiturbine is a pure expansion engine, while the Wankel and most other rotary engines are not, it is well-suited as a compressed fluid engine, air engine or air motor.

Regusci

Armando Regusci's version of the air engine couples the transmission system directly to the wheel, and has variable torque from zero to the maximum, enhancing efficiency. Regusci's patents date from 1990.

Team Psycho-Active

Psycho-Active is developing a multi-fuel/air-hybrid chassis which is intended to serve as the foundation for a line of automobiles. Claimed performance is 50 hp/litre.

Defunct Air Engine Designs

Conger Motor

Milton M. Conger in 1881 patented and supposedly built a motor that ran off compressed air or steam that using a flexible tubing which will form a wedge-shaped or inclined wall or abutment in the rear of the tangential bearing of the wheel, and propel it with greater or less speed according to the pressure of the propelling medium.

Cars

Since about 1990 several companies have claimed to be developing compressed air cars, but none are available. Typically the main claimed advantages are: no roadside pollution, low cost, use of cooking oil for lubrication, and integrated air conditioning.

The time required to refill a depleted tank is important for vehicle applications. "Volume transfer" moves pre-compressed air from a stationary tank to the vehicle tank almost instantaneously. Alternatively, a stationary or on-board compressor can compress air on demand, possibly requiring several hours. The cost of driving such car is typically projected to be around €0.75 per 100 kilometres (62 mi) with a complete refill at a service station costing about US$3.

Types of Systems

Cryogenic Systems

A special CAES system has been created that uses liquid air as an energy carrier. This system is called Highview Power Storage's CryoEnergy System (CES).

Hybrid Systems

Brayton cycle engines compress and heat air with a fuel suitable for an internal combustion engine. For example, natural gas or biogas heat compressed air, and then a conventional gas turbine engine or the rear portion of a jet engine expands it to produce work.

Compressed air engines can recharge an electric battery. The apparently defunct Energine promoted its Pne-PHEV or Pneumatic Plug-in Hybrid Electric Vehicle-system).

Existing Hybrid Systems

Huntorf, Germany in 1978, and McIntosh, Alabama in 1991 (USA) commissioned hybrid power plants. Both systems use off-peak energy for air compression. The McIntosh plant achieves its 24-hour operating cycle by burning a natural gas/compressed air mix.

Future Hybrid Systems

The Iowa Stored Energy Park (ISEP) will use aquifer storage rather than cavern storage. The displacement of water in the aquifer results in regulation of the air pressure by the constant hydrostatic pressure of the water. A spokesperson for ISEP claims, "you can optimise your equipment for better efficiency if you have a constant pressure." Power output of the McIntosh and Iowa systems is in the range of 2–300 MW. Additional facilities are under development in Norton, Ohio. First Energy, an Akron, Ohio electric utility obtained development rights to the 2,700 MW Norton project in November, 2009.

Lake or Ocean Storage

Deep water in lakes and the ocean can provide pressure without requiring high-pressure vessels or drilling into salt caverns or aquifers. The air goes into inexpensive, flexible containers such as plastic bags below in deep lakes or off sea coasts with steep drop-offs. Obstacles include the limited number of suitable locations and the need for high-pressure pipelines between the surface and the containers. Since the containers would be very inexpensive, the need for great pressure (and great depth) may not be as important. A key benefit of systems built on this concept is that charge and discharge pressures are a constant function of depth. Carnot inefficiencies can thereby be reduced in the power plant. Carnot efficiency can be increased by using multiple charge and discharge stages and using inexpensive heat sources and sinks such as cold water from rivers or hot water from solar ponds. Ideally, the system must be very clever—for example, by cooling air before pumping on summer days. It must be engineered to avoid inefficiency, such as wasteful pressure changes caused by inadequate piping diametre.

A nearly isobaric solution is possible if the compressed gas is used to drive a hydroelectric system. However, this solution requires large pressure tanks located on land (as well as the underwater air bags). Also, hydrogen gas is the preferred fluid, since other gases suffer from substantial hydrostatic pressures at even relatively modest depths (such as 500 metres). The University of Nottingham is one centre of research on seabed–anchored energy bags. E.ON, one of Europe's leading power and gas companies, has provided €1.4 million (£1.1 million) in funding to develop undersea air storage bags. Hydrostor in Canada is developing a commercial system of underwater storage "accumulators" for compressed air energy storage, starting at the 1 to 4 MW scale.

Radial Turbine

A Radial turbine is a turbine in which the flow of the working fluid is radial to the shaft. The difference between axial and radial turbines consists in the way the air flows through the components (compressor and turbine). Whereas for an axial turbine the rotor is 'impacted' by the air flow, for a radial turbine, the flow is smoothly orientated at 90 degrees by the compressor towards the combustion chamber and driving the turbine in the same way water drives a watermill. The result is less mechanical and thermal stress which enables a radial turbine to be simpler, more robust and more efficient (in a similar power range as axial turbines). When it comes to high power ranges (above 5 MW) the radial turbine is no longer competitive (heavy and expensive rotor) and the efficiency becomes similar to that of the axial turbines.

Nikola Tesla's Bladeless Radial Turbine

In the early 1900s, Nikola Tesla developed and patented his Bladeless Turbine. One of the difficulties with bladed turbines is the complex and highly precise requirements for balancing and manufacturing the bladed rotor which has to be very well balanced. The blades are subject to corrosion and cavitation. Tesla attacked this problem by substituting a series of closely spaced disks for the blades of the rotor. The working fluid flows between the disks and transfers its energy to the rotor by means of the boundary layer effect or adhesion and viscosity rather than by impulse or reaction. Tesla turbines are typically most effective at small diametres, so they are not seen powering jet aircraft or large power plants, but for microturbine applications they have ease of manufacture and good efficiency, two characteristics that are not found in most other turbine configurations.

Euler's Turbine Equation

Euler's Turbine Equation states that

$$W = (U2\text{-}U1)+(V2\text{-}V1)+(W2\text{-}W1),$$

where U = Blade Velocity = r*omega (omega - Angular Velocity)

V = Absolute Velocity

W = Relative Velocity

For an axial machine, you have Zero Degree of Reaction that makes U2-U1 = 0. This is one reason for the fact that axial turbines have low power output. For a radial turbine, along the flowpath, Radius reduces and hence U2 - U1 is not equal to zero, which proves that for similar turbines, radial turbine has higher power output.

Advantages Compared to Axial Turbines

Thanks to lower thermal and mechanical stress on the turbine tips, it is possible to boost power quite significantly by increasing the turbine entry temperature (increasing fuel input) which results in an improved mechanical efficiency. The lower mechanical stresses also enable radial turbines to handle single stage compression and expansion.

As a result, the radial turbine does not need to be air cooled, which means that all the air entering the compressor is used only to drive the turbine which gives the radial design a strong advantage for co-generation applications. Another result of avoiding air cooling is that power and efficiency are kept almost constant during the lifetime of the radial turbine whereas an axial gas turbine needs to be washed often to maintain ISO performance standards. The other advantage of such a simple rotor is that the bearings can be placed at the front, in the cold part, so less lubrication oil is needed, and there are no thermal losses due to lubrication of the hot parts of the rotor.

Chapter 6

Jet Engine

A jet engine is a reaction engine that discharges a fast moving jet to generate thrust by *jet propulsion* and in accordance with Newton's laws of motion. This broad definition of jet engines includes turbojets, turbofans, rockets, ramjets, pulse jets. In general, most jet engines are internal combustion engines but non-combusting forms also exist.

In common parlance, the term *jet engine* loosely refers to an internal combustion airbreathing jet engine (a *duct engine*). These typically consist of an engine with a rotary (rotating) air compressor powered by a turbine ("Brayton cycle"), with the leftover power providing thrust via a propelling nozzle. These types of jet engines are primarily used by jet aircraft for long distance travel. Early jet aircraft used turbojet engines which were relatively inefficient for subsonic flight. Modern subsonic jet aircraft usually use high-bypass turbofan engines which give high speeds, as well as (over long distances) fuel efficiency that is about as good as piston and propeller aeroengines.

History of the Jet Engine

The jet engine has a long history, from early steam devices in the 2nd century BCE to the modern turbofans and scramjets.

Precursors

Jet engines can be dated back to the invention of the aeolipile around 150 BCE. This device used steam power directed through two nozzles so as to cause a sphere to spin rapidly on its axis. So far as is known, it was not used for supplying mechanical power, and the potential practical applications of this invention were not recognised. It was simply considered a curiosity.

Jet propulsion only literally and figuratively took off with the invention of the rocket by the Chinese in the 13th century fireworks but gradually progressed to propel formidable weaponry; and there the technology stalled for hundreds of years.

Archytas, the founder of mathematical mechanics, as described in the writings of Aulus Gellius five centuries after him, was reputed to have designed and built the first artificial, self-propelled flying device. This device was a bird-shaped model propelled by a jet of what was probably steam, said to have actually flown some 200 metres.

Ottoman Lagari Hasan Celebi is said to have taken off in 1633 with what was described to be a cone-shaped rocket and then to have glided with wings into a successful landing, winning a position in the Ottoman army. However, this was essentially a stunt. The problem was that rockets are simply too inefficient at low speeds to be useful for general aviation.

The earliest attempts at airbreathing jet engines were hybrid designs in which an external power source first compressed air, which was then mixed with fuel and burned for jet thrust. In one such system, called a *thermojet* by Secondo Campini but more commonly, motorjet, the air was compressed by a fan driven by a conventional piston engine. Examples include the Caproni Campini N.1 and the Japanese Tsu-11 engine intended to power Ohka kamikaze planes towards the end of World War II. None were entirely successful and the CC.2 ended up being slower than the same design with a traditional engine and propeller combination.

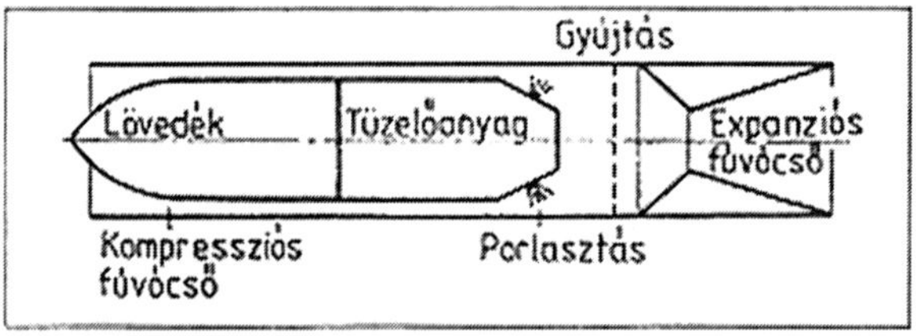

Figure 1: Albert Fono's ramjet-cannonball from 1915

In 1913 Rene Lorin came up with a form of jet engine, the subsonic pulsejet, which would have been somewhat more efficient, but he had no way to achieve high enough speeds for it to operate, and the concept remained theoretical for quite some time.

Even before the start of World War II, engineers were beginning to realise that the piston engine was self-limiting in terms of the maximum performance which could be attained; the limit was due to issues related to propulsive efficiency, which declined as blade tips approached the speed of sound. If engine, and thus aircraft, performance were ever to increase beyond such a barrier, a way would have to be found to radically improve the design of the piston engine, or a wholly new type of powerplant would have to be developed. This was the motivation behind the development of the gas turbine engine, commonly called a "jet" engine, which would become almost as revolutionary to aviation as the Wright brothers' first flight.

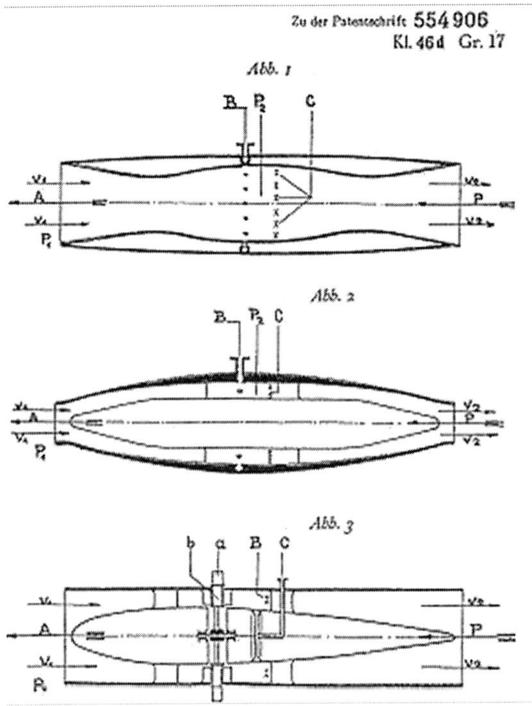

Figure 2: Albert Fono's German patent for jet Engines (January 1928- granted 1932). The third illustration is a turbojet

The key to a practical jet engine was the gas turbine, used to extract energy from the engine itself to drive the compressor. The gas turbine was not an idea developed in the 1930s: the patent for a stationary turbine was granted to John Barber in England in 1791. The first gas turbine to successfully run self-sustaining was built in 1903 by Norwegian engineer Ægidius Elling. Limitations in design and practical engineering and metallurgy prevented such engines reaching manufacture. The main problems were safety, reliability, weight and, especially, sustained operation.

In Hungary, Albert Fono in 1915 devised a solution for increasing the range of artillery, comprising a gun-launched projectile which was to be united with a ramjet propulsion unit. This was to make it possible to obtain a long range with low initial muzzle velocities, allowing heavy shells to be fired from relatively lightweight guns. Fono submitted his invention to the Austro-Hungarian Army but the proposal was rejected. In 1928 he applied for a German patent on aircraft powered by supersonic ramjets, and this was awarded in 1932.

The first patent for using a gas turbine to power an aircraft was filed in 1921 by Frenchman Maxime Guillaume. His engine was an axial-flow turbojet.

In 1923, Edgar Buckingham of the US National Bureau of Standard published a report expressing scepticism that jet engines would be economically competitive with prop driven aircraft at the low altitudes and airspeeds of the period: "there does not appear to be, at present, any prospect whatever that jet propulsion of the sort here considered will ever be of practical value, even for military purposes."

Instead, by the 1930s, the piston engine in its many different forms (rotary and static radial, aircooled and liquid-cooled inline) was the only type of powerplant available to aircraft designers. This was acceptable as long as only low performance aircraft were required, and indeed all that were available.

Pre World War II

Figure 3: The Whittle W.2/700 engine flew in the Gloster E.28/39, the first British aircraft to fly with a turbojet engine, and the Gloster Meteor

In 1928, RAF College Cranwell cadet Frank Whittle formally submitted his ideas for a turbo-jet to his superiors. In October 1929 he developed his ideas further. On 16 January 1930 in England, Whittle submitted his first patent (granted in 1932). The patent showed a two-stage axial compressor feeding a single-sided centrifugal compressor. Practical axial compressors were made possible by ideas from A.A.Griffith in a seminal paper in 1926 ("An Aerodynamic Theory of Turbine Design"). Whittle would later concentrate on the simpler centrifugal compressor only, for a variety of practical reasons. Whittle had his first engine running in April 1937. It was liquid-fuelled, and included a self-contained fuel pump. Whittle's team experienced near-panic when the engine would not stop, accelerating even after the fuel was switched off.

It turned out that fuel had leaked into the engine and accumulated in pools. So the engine would not stop until all the leaked fuel had burned off. Whittle was unable to interest the government in his invention, and development continued at a slow pace.

In 1935 Hans von Ohain started work on a similar design in Germany, apparently unaware of Whittle's work. His first device was strictly experimental and could only run under external power, but he was able to demonstrate the basic concept.

Ohain was then introduced to Ernst Heinkel, one of the larger aircraft industrialists of the day, who immediately saw the promise of the design. Heinkel had recently purchased the Hirth engine company, and Ohain and his master machinist Max Hahn were set up there as a new division of the Hirth company.

They had their first HeS 1 centrifugal engine running by September 1937. Unlike Whittle's design, Ohain used hydrogen as fuel, supplied under external pressure. Their subsequent designs culminated in the gasoline-fuelled HeS 3 of 1,100 lbf (5 kN), which was fitted to Heinkel's simple and compact He 178 airframe and flown by Erich Warsitz in the early morning of August 27, 1939, from Rostock-Marienehe aerodrome, an impressively short time for development. The He 178 was the world's first jet plane.

The world's first turboprop was the Jendrassik Cs-1 designed by the Hungarian mechanical engineer Gyorgy Jendrassik. It was produced and tested in the Ganz factory in Budapest between 1938 and 1942. It was planned to fit to the Varga RMI-1 X/H twin-engined reconnaissance bomber designed by Lászlo Varga in 1940, but the program was cancelled. Jendrassik had also designed a small-scale 75 kW turboprop in 1937.

Whittle's engine was starting to look useful, and his Power Jets Ltd. started receiving Air Ministry money. In 1941 a flyable version of the engine called the W.1, capable of 1000 lbf (4 kN) of thrust, was fitted to the Gloster E28/39 airframe specially built for it, and first flew on May 15, 1941 at RAF Cranwell.

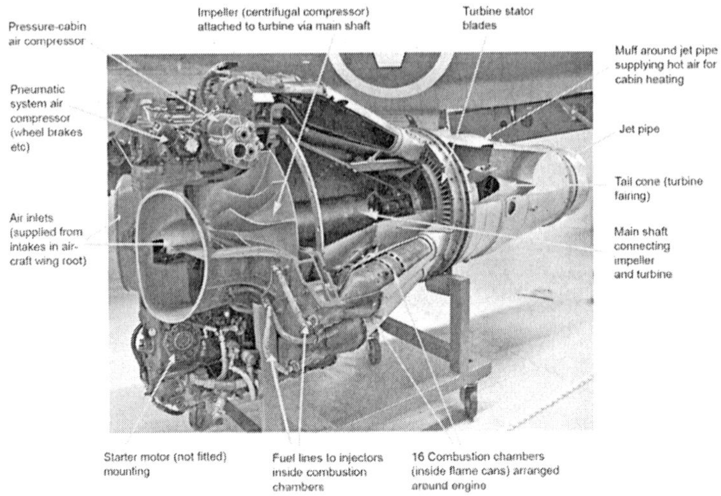

Figure 4: A picture of an early centrifugal engine (DH Goblin II) sectioned to show its internal components

A Scottish aircraft engine designer, Frank Halford, working from Whittle's ideas developed a "straight through" version of the centrifugal jet; his design became the de Havilland Goblin.

One problem with both of these early designs, which are called centrifugal-flow engines, was that the compressor worked by "throwing" (accelerating) air outward from the central intake to the outer periphery of the engine, where the air was then compressed by a divergent duct setup, converting its velocity into pressure.

An advantage of this design was that it was already well understood, having been implemented in centrifugal superchargers, then in widespread use on piston engines. However, given the early technological limitations on the shaft speed of the engine, the compressor needed to have a very large diametre to produce the power required.

This meant that the engines had a large frontal area, which made it less useful as an aircraft powerplant due to drag. A further disadvantage was that the air flow had to be "bent" to flow rearwards through the combustion section and to the turbine and tailpipe, adding complexity and lowering efficiency.

Nevertheless, these types of engines had the major advantages of light weight, simplicity and reliability, and development rapidly progressed to practical airworthy designs.

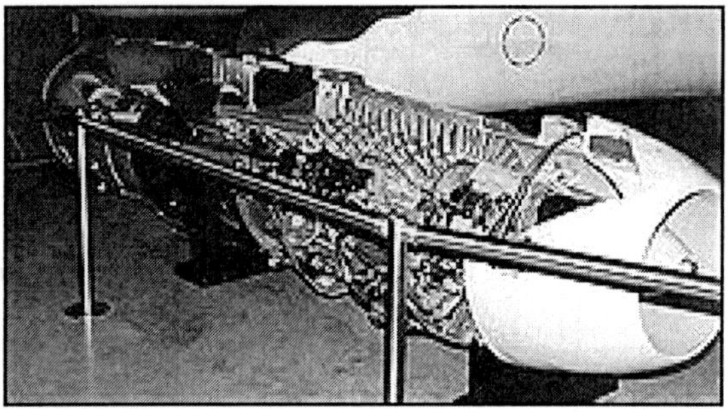

Figure 5: A cutaway of the Junkers Jumo 004 engine

Austrian Anselm Franz of Junkers' engine division (*Junkers Motoren* or Jumo) addressed these problems with the introduction of the axial-flow compressor. Essentially, this is a turbine in reverse. Air coming in the front of the engine is blown towards the rear of the engine by a fan stage (convergent ducts), where it is crushed against a set of non-rotating blades called *stators* (divergent ducts). The process is nowhere near as powerful as the centrifugal compressor, so a number of these pairs of fans and stators are placed in series to get the needed compression. Even with all the added complexity, the resulting engine is much smaller in diametre and thus, more aerodynamic.

Jumo was assigned the next engine number in the RLM numbering sequence, 4, and the result was the Jumo 004 engine. After many lesser technical difficulties were solved, mass production of this engine started in 1944 as a powerplant for the world's first jet-fighter aircraft, the Messerschmitt Me 262. A variety of reasons conspired to delay the engine's availability, this delay caused the fighter to arrive too late to decisively impact Germany's position in World War II. Nonetheless, it will be remembered as the first use of jet engines in service.

In the UK, their first axial-flow engine, the Metrovick F.2, ran in 1941 and was first flown in 1943. Although more powerful than the centrifugal designs at the time, the Ministry considered its complexity and unreliability a drawback in wartime. The work at Metrovick led to the Armstrong Siddeley Sapphire engine which would be built in the US as the J65.

Post World War II

Following the end of the war the German jet aircraft and jet engines were extensively studied by the victorious allies and contributed to work on early Soviet and US jet fighters. The legacy of the axial-flow engine is seen in the fact that practically all jet engines on fixed wing aircraft have had some inspiration from this design.

Centrifugal-flow engines have improved since their introduction. With improvements in bearing technology the shaft speed of the engine was increased, greatly reducing the diametre of the centrifugal compressor. The short engine length remains an advantage of this design, particularly for use in helicopters where overall size is more important than frontal area. Also as their engine components are more robust they are less liable to foreign object damage than axial-flow compressor engines.

Although German designs were more advanced aerodynamically, the combination of simplicity and the lack of requisite rare metals for the necessary advanced metallurgy (such as tungsten, chromium and titanium) for high-stress components such as turbine blades and bearings, etc.) meant that the later produced German engines had a short service life and had to be changed after 10–25 hours. British engines were also widely manufactured under license in the US, and were sold to Soviet Russia who reverse engineered them with the Nene going on to power the famous MiG-15. American and Soviet designs, independent axial-flow types for the most part, would strive to attain superior performance until the 1960s, although the General Electric J47 provided excellent service in the F-86 Sabre in the 1950s.

By the 1950s the jet engine was almost universal in combat aircraft, with the exception of cargo, liaison and other speciality types. By this point some of the British designs were already cleared for civilian use, and had appeared on early models like the de Havilland Comet and Avro Canada Jetliner. By the 1960s all large civilian aircraft were also jet powered, leaving the piston engine in such low-cost niche roles such as cargo flights.

Relentless improvements in the turboprop pushed the piston engine (an internal combustion engine) out of the mainstream entirely, leaving it serving only the smallest general aviation designs and some use in drone aircraft. The ascension of the jet engine to almost universal use in aircraft took well under twenty years.

However, the story was not quite at an end, for the efficiency of turbojet engines was still rather worse than piston engines, but by the 1970s with the advent of high bypass jet engines, an innovation not foreseen by the early commentators like Edgar Buckingham, at high speeds and high altitudes that seemed absurd to them, only then did the fuel efficiency finally exceed that of the best piston and propeller engines, and the dream of fast, safe, economical travel around the world finally arrived, and their dour, if well founded for the time, predictions that jet engines would never amount to much, were killed forever.

Uses

Figure 6: A JT9D turbofan jet engine undergoing maintenance on a Boeing 747 aircraft

Jet engines are usually used as aircraft engines for jet aircraft. They are also used for cruise missiles and unmanned aerial vehicles.

In the form of rocket engines they are used for fireworks, model rocketry, spaceflight, and military missiles. Jet engines have also been used to propel high speed cars, particularly drag racers, with the all time record held by a rocket car. A turbofan powered car ThrustSSC currently holds the land speed record.

Jet engine designs are frequently modified for non-aircraft applications, as industrial gas turbines. These are used in electrical power generation, for powering water, natural gas, or oil pumps, and providing propulsion for ships and locomotives. Industrial gas turbines can create up to 50,000 shaft horsepower. Many of these engines are derived from older military turbojets such as the Pratt & Whitney J57 and J75 models. There is also a derivative of the P&W JT8D low-bypass turbofan that creates up to 35,000 HP.

Types

There are a large number of different types of jet engines, all of which achieve forward thrust from the principle of *jet propulsion.*

Airbreathing Jet Engine

An airbreathing jet engine (or *ducted jet engine*) is a jet engine propelled by a jet of hot exhaust gases formed from air that is drawn into the engine via an inlet duct.

All practical airbreathing jet engines are internal combustion engines that directly heat the air by combusting fuel, with the resultant hot gases used for propulsion via a propulsive nozzle, although other techniques for heating the air have been experimented with. Most jet engines are turbofans and some are turbojets which use gas turbines to give high pressure ratios and are able to get high efficiency, but a few use simple ram effect or pulse combustion to give compression. Most commercial aircraft possess turbofans, these have an enlarged air compressor which permit them to generate most of their thrust from air which bypasses the combustion chamber. Airbreathing jet engines are mostly used for powering jet aircraft, but have seen rare other uses such as jet cars.

Types of Airbreathing Jet Engines

Airbreathing jet engines are nearly always internal combustion engines that obtain propulsion from the combustion of fuel inside the engine. Oxygen present in the atmosphere is used to oxidise a fuel source, typically a hydrocarbon-based jet fuel. The burning mixture expands greatly in volume, driving heated air through a propelling nozzle.

Gas turbine powered engines:

- turbojet
- turbofan.

Ram powered jet engine:

- ramjet
- scramjet.

Pulsed combustion jet engine:

- pulse detonation engine
- pulse jet engine
- motorjet.

Turbojet engine

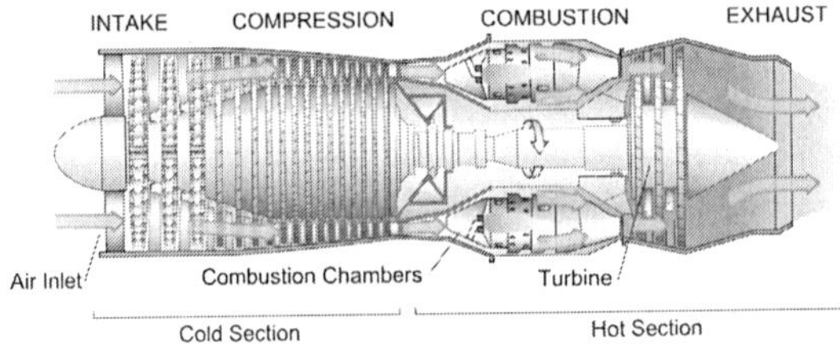

Figure 7: Turbojet engine layout

The turbojet is the oldest kind of general-purpose jet engine. Two engineers, Frank Whittle in the United Kingdom and Hans von Ohain in Germany, developed the concept independently into practical engines during the late 1930s.

Turbojets consist of an air inlet, an air compressor, a combustion chamber, a gas turbine (that drives the air compressor) and a nozzle. The air is compressed into the chamber, heated and expanded by the fuel combustion and then allowed to expand out through the turbine into the nozzle where it is accelerated to high speed to provide propulsion.

Turbojets are quite inefficient if flown below about Mach 2, and very noisy. Most modern aircraft use turbofans instead for economic reasons. Turbojets are still very common in medium range cruise missiles, due to their high exhaust speed, low frontal area and relative simplicity.

Turbofan Engine

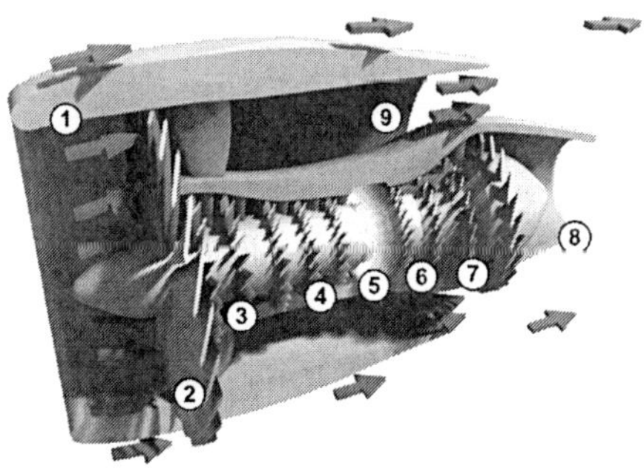

Figure 8: An animated turbofan engine

Most modern jet engines are actually turbofans, where the low pressure compressor acts as a fan, supplying supercharged air not only to the engine core, but to a bypass duct. The bypass airflow either passes to a separate 'cold nozzle' or mixes with low pressure turbine exhaust gases, before expanding through a 'mixed flow nozzle'.

In the 1960s there was little difference between civil and military jet engines, apart from the use of afterburning in some (supersonic) applications. Today, turbofans are used for airliners because they give an exhaust speed that is better matched for subsonic airliners. At airliner flight speeds, conventional turbojet engines generate an exhaust that ends up travelling very fast backwards (rocket engines are worse still), and this wastes energy. By emitting the exhaust so that it ends up travelling more slowly, better fuel consumption is achieved as well as higher thrust at low speeds. In addition, the lower exhaust speed gives much lower noise.

Thus civil turbofans today have a low exhaust speed (low *specific thrust* -net thrust divided by airflow) to keep jet noise to a minimum and to improve fuel efficiency. Consequently the bypass ratio (bypass flow divided by core flow) is relatively high (ratios from 4:1 up to 8:1 are common). Only a single fan stage is required, because a low specific thrust implies a low fan pressure ratio.

Military turbofans, however, have a relatively high specific thrust, to maximise the thrust for a given frontal area, jet noise being of less concern in military uses relative to civil uses. Multistage fans are normally needed to reach the relatively high fan pressure ratio needed for high specific thrust. Although high turbine inlet temperatures are often employed, the bypass ratio tends to be low, usually significantly less than 2.0.

Major Components

The major components of a jet engine are similar across the major different types of engines, although not all engine types have all components. The major parts include:

- Cold Section:
  - Air intake (Inlet)—For subsonic aircraft, the air intake to a jet engine consists essentially of an opening which is designed to minimise drag. The air reaching the compressor of a normal jet engine must be travelling below the speed of sound, even for supersonic aircraft, to allow smooth flow through compressor and turbine blades. At supersonic flight speeds, shockwaves form in the intake system. These help compress the air, but also there is inevitable reduction in the recovered pressure at inlet to the compressor. Some supersonic intakes use devices such as a cone or a ramp to increase pressure recovery.
  - Compressor or Fan—The compressor is made up of stages. Each stage consists of vanes which rotate, and stators which remain stationary. As air is drawn deeper through the compressor, its heat and pressure increases. Energy is derived from the turbine, passed along the shaft.

  - o Bypass ducts—Much of the thrust of essentially all modern jet engines comes from air from the front compressor that bypasses the combustion chamber and gas turbine section that leads directly to the nozzle or afterburner (where fitted).
- Common:
  - o Shaft—The shaft connects the turbine to the compressor, and runs most of the length of the engine. There may be as many as three concentric shafts, rotating at independent speeds, with as many sets of turbines and compressors. Other services, like a bleed of cool air, may also run down the shaft.
- Diffuser section: - This section is a divergent duct that utilises Bernoulli's principle to decrease the velocity of the compressed air to allow for easier ignition. And, at the same time, continuing to increase the air pressure before it enters the combustion chamber.
- Hot section:

o Combustor or Can or Flameholders or Combustion Chamber—This is a chamber where fuel is continuously burned in the compressed air.

o Turbine—The turbine is a series of bladed discs that act like a windmill, gaining energy from the hot gases leaving the combustor. Some of this energy is used to drive the compressor, and in some turbine engines (i.e. turboprop, turboshaft or turbofan engines), energy is extracted by additional turbine discs and used to drive devices such as propellers, bypass fans or helicopter rotors. One type, a free turbine, is configured such that the turbine disc driving the compressor rotates independently of the discs that power the external components. Relatively cool air, bled from the compressor, may be used to cool the turbine blades and vanes, to prevent them from melting.

o Afterburner or reheat (chiefly UK)—(mainly military) Produces extra thrust by burning extra fuel, usually inefficiently, to significantly raise Nozzle Entry Temperature at the exhaust. Owing to a larger volume flow (i.e. lower density) at exit from the afterburner, an increased nozzle flow area is required, to maintain satisfactory engine matching, when the afterburner is alight.

- o Exhaust or Nozzle—Hot gases leaving the engine exhaust expand to atmospheric pressure via a nozzle, the objective being to produce a high velocity jet. In most cases, the nozzle is convergent and of fixed flow area. If the Nozzle Pressure Ratio (Nozzle Entry Pressure/Ambient Pressure) is very high, to maximise thrust it may be worthwhile, despite the additional weight, to fit a convergent-divergent (de Laval) nozzle- a supersonic nozzle. As the name suggests, initially this type of nozzle is convergent, but beyond the throat (smallest flow area), the flow area starts to increase to form the divergent portion. The expansion to atmospheric pressure and supersonic gas velocity continues downstream of the throat, whereas in a convergent nozzle the expansion beyond sonic velocity occurs externally, in the exhaust plume. The former process is more efficient than the latter.

The various components named above have constraints on how they are put together to generate the most efficiency or performance. The performance and efficiency of an engine can never be taken in isolation; for example fuel/distance efficiency of a supersonic jet engine maximises at about Mach 2, whereas the drag for the vehicle carrying it is increasing as a square law and has much extra drag in the transonic region. The highest fuel efficiency for the overall vehicle is thus typically at Mach ~0.85. For the engine optimisation for its intended use, important here is air intake design, overall size, number of compressor stages (sets of blades), fuel type, number of exhaust stages, metallurgy of components, amount of bypass air used, where the bypass air is introduced, and many other factors. For instance, let us consider design of the air intake.

Operation—Engine Cycle

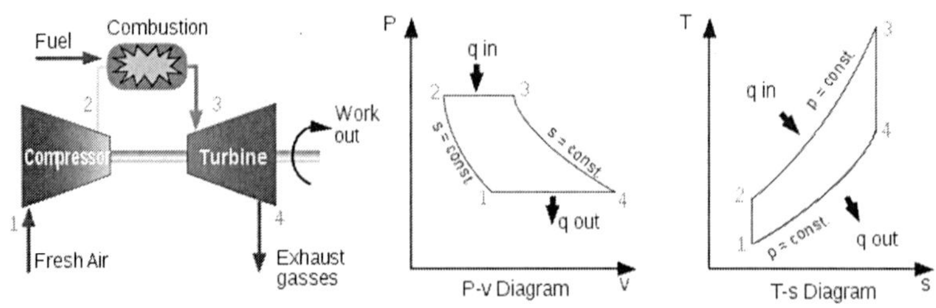

Figure 9: Brayton cycle

The thermodynamics of a typical air-breathing jet engine are modelled approximately by a Brayton Cycle which is a thermodynamic cycle that describes the workings of the gas turbine engine, basis of the airbreathing jet engine and others. It is named after George Brayton (1830–1892), the American engineer who developed it, although it was originally proposed and patented by Englishman John Barber in 1791. It is also sometimes known as the Joule cycle.

Thrust Lapse

The nominal net thrust quoted for a jet engine usually refers to the Sea Level Static (SLS) condition, either for the International Standard Atmosphere (ISA) or a hot day condition (e.g. ISA+10 °C). As an example, the GE90-76B has a take-off static thrust of 76,000 lbf (360 kN) at SLS, ISA+15 °C. Naturally, net thrust will decrease with altitude, because of the lower air density. There is also, however, a flight speed effect. Initially as the aircraft gains speed down the runway, there will be little increase in nozzle pressure and temperature, because the ram rise in the intake is very small. There will also be little change in mass flow. Consequently, nozzle gross thrust initially only increases marginally with flight speed. However, being an air breathing engine (unlike a conventional rocket) there is a penalty for taking on-board air from the atmosphere. This is known as ram drag. Although the penalty is zero at static conditions, it rapidly increases with flight speed causing the net thrust to be eroded.

As flight speed builds up after take-off, the ram rise in the intake starts to have a significant effect upon nozzle pressure/temperature and intake airflow, causing nozzle gross thrust to climb more rapidly. This term now starts to offset the still increasing ram drag, eventually causing net thrust to start to increase. In some engines, the net thrust at say Mach 1.0, sea level can even be slightly greater than the static thrust. Above Mach 1.0, with a subsonic inlet design, shock losses tend to decrease net thrust, however a suitably designed supersonic inlet can give a lower reduction in intake pressure recovery, allowing net thrust to continue to climb in the supersonic regime.

Safety and Reliability

Jet engines are usually very reliable and have a very good safety record. However, failures do sometimes occur.

Engine Surge

In some cases in jet engines the conditions in the engine due to airflow entering the engine or other variations can cause the compressor blades to stall. When this occurs the pressure in the engine blows out past the blades, and the stall is maintained until the pressure has decreased, and the engine has lost all thrust. The compressor blades will then usually come out of stall, and re-pressurise the engine. If conditions are not corrected, the cycle will usually repeat. This is called surge. Depending on the engine this can be highly damaging to the engine and creates worrying vibrations for the crew.

Compressor Blade Containment

Due to "foreign object damage"- material being sucked into the engine- the most likely failure is often compressor blade failure, and modern jet engines are designed with structures that can catch these blades and keep them contained within the engine casing. Verification of a jet engine design involves testing that this system works correctly.

Bird Strike

Bird strike is an aviation term for a collision between a bird and an aircraft. It is a common threat to aircraft safety and has caused a number of fatal accidents. In 1988 an Ethiopian Airlines Boeing 737 sucked pigeons into both engines during take-off and then crashed in an attempt to return to the Bahir Dar airport; of the 104 people aboard, 35 died and 21 were injured. In another incident in 1995, a Dassault Falcon 20 crashed at a Paris airport during an emergency landing attempt after sucking lapwings into an engine, which caused an engine failure and a fire in the aeroplane fuselage; all 10 people on board were killed. In 2009, on US Airways Flight 1549, a Airbus A320 aircraft sucked in one bird in each engine. The plane landed in the Hudson River after taking off from LaGuardia International Airport in New York City. There were no fatalities.

Modern jet engines have the capability of surviving an ingestion of a bird. Small fast planes, such as military jet fighters, are at higher risk than big heavy multi-engine ones. This is due to the fact that the fan of a high-bypass turbofan engine, typical on transport aircraft, acts as a centrifugal separator to force ingested materials (birds, ice, etc.) to the outside of the fan's disc. As a result, such materials go through the relatively unobstructed bypass duct, rather than through the core of the engine, which contains the smaller and more delicate compressor blades. Military aircraft designed for high-speed flight typically have pure turbojet, or low-bypass turbofan engines, increasing the risk that ingested materials will get into the core of the engine to cause damage.

The highest risk of the bird strike is during the takeoff and landing, in low altitudes, which is in the vicinity of the airports.

Volcanic Ash

If a jet plane is flying through air densely contaminated with volcanic ash, there is risk of ingested ash eroding the front blades, melting in the combustion heat, and re-freezing sticking to the rear blades, affecting performance and perhaps stopping the engine; as well as triggering long-term corrosion.

Uncontained Failures

One class of failures that has caused accidents in particular is uncontained failures, where rotary parts of the engine break off and exit through the case. These can cut fuel or control lines, and can penetrate the cabin.

Although fuel and control lines are usually duplicated for reliability, the crash of United Airlines Flight 232 was caused when hydraulic fluid lines for all three independent hydraulic systems were simultaneously severed by shrapnel from an uncontained engine failure.

Prior to the United 232 crash, the probability of a simultaneous failure of all three hydraulic systems was considered as high as a billion-to-one. However, the statistical models used to come up with this figure did not account for the fact that the number-two engine was mounted at the tail close to all the hydraulic lines, nor the possibility that an engine failure would release many fragments in many directions.

Since then, more modern aircraft engine designs have focused on keeping shrapnel from penetrating the cowling or ductwork, and have increasingly utilised high-strength composite materials to achieve the required penetration resistance while keeping the weight low.

Economic Considerations

In 2007, the cost of jet fuel, while highly variable from one airline to another, averaged 26.5% of total operating costs, making it the single largest operating expense for most airlines.

Environmental Considerations

Jet engines are usually run on fossil fuel propellant, and are thus a source of carbon dioxide in the atmosphere. Jet engines can use biofuels or hydrogen, although the production of the latter is usually made from fossil fuels.

About 7.2% of the oil used in 2004 was consumed by jet engines.

Some scientists believe that jet engines are also a source of global dimming due to the water vapour in the exhaust causing cloud formations.

Nitrogen compounds are also formed from the combustion process from atmospheric nitrogen. At low altitudes this is not thought to be especially harmful, but for supersonic aircraft that fly in the stratosphere some destruction of ozone may occur.

Sulphates are also emitted if the fuel contains sulphur.

Advanced Designs

Ramjet

A ramjet is a form of airbreathing jet engine using the engine's forward motion to compress incoming air, without a rotary compressor. Ramjets cannot produce thrust at zero airspeed and thus cannot move an aircraft from a standstill. Ramjets require considerable forward speed to operate well, and as a class work most efficiently at speeds around Mach 3. This type of jet can operate up to speeds of Mach 6.

Ramjets can be particularly useful in applications requiring a small and simple engine for high speed use, such as missiles, while weapon designers are looking to use ramjet technology in artillery shells to give added range: it is anticipated that a 120-mm mortar shell, if assisted by a ramjet, could attain a range of 22 mi (35 km). They have also been used successfully, though not efficiently, as tip jets on helicopter rotors.

Ramjets are frequently confused with pulsejets, which use an intermittent combustion, but ramjets employ a continuous combustion process, and are a quite distinct type of jet engine.

J-58 Combined Ramjet/Turbojet

The SR-71 Blackbird's Pratt & Whitney J58 engines were rather unusual. They could convert in flight from being largely a turbojet to being largely a compressor-assisted ramjet.

At high speeds (above Mach 2.4), the engine used variable geometry vanes to direct excess air through 6 bypass pipes from downstream of the fourth compressor stage into the afterburner. 80% of the SR-71's thrust at high speed was generated in this way, giving much higher thrust, improving specific impulse by 10-15%, and permitting continuous operation at Mach 3.2. The name coined for this setup is *turbo-ramjet*.

Hydrogen Fuelled Air-breathing Jet Engines

Jet engines can be run on almost any fuel. Hydrogen is a highly desirable fuel, as, although the energy per mole is not unusually high, the molecule is very much lighter than other molecules. The energy per kg of hydrogen is twice that of more common fuels and this gives twice the specific impulse. In addition, jet engines running on hydrogen are quite easy to build—the first ever turbojet was run on hydrogen. Also, although not duct engines, hydrogen-fuelled rocket engines have seen extensive use.

However, in almost every other way, hydrogen is problematic. The downside of hydrogen is its density; in gaseous form the tanks are impractical for flight, but even in the form of liquid hydrogen it has a density one fourteenth that of water. It is also deeply cryogenic and requires very significant insulation that precludes it being stored in wings. The overall vehicle would end up being very large, and difficult for most airports to accommodate. Finally, pure hydrogen is not found in nature, and must be manufactured either via steam reforming or expensive electrolysis. Nevertheless, research is ongoing and hydrogen-fuelled aircraft designs do exist that may be feasible.

Precooled Jet Engines

An idea originated by Robert P. Carmichael in 1955 is that hydrogen-fuelled engines could theoretically have much higher performance than hydrocarbon-fuelled engines if a heat exchanger were used to cool the incoming air. The low temperature allows lighter materials to be used, a higher mass-flow through the engines, and permits combustors to inject more fuel without overheating the engine.

This idea leads to plausible designs like Reaction Engines SABRE, that might permit single-stage-to-orbit launch vehicles, and ATREX, which could permit jet engines to be used up to hypersonic speeds and high altitudes for boosters for launch vehicles. The idea is also being researched by the EU for a concept to achieve non-stop antipodal supersonic passenger travel at Mach 5 (Reaction Engines A2).

Nuclear-powered Ramjet

Project Pluto was a nuclear-powered ramjet, intended for use in a cruise missile. Rather than combusting fuel as in regular jet engines, air was heated using a high-temperature, unshielded nuclear reactor. This dramatically increased the engine burn time, and the ramjet was predicted to be able to cover any required distance at supersonic speeds (Mach 3 at tree-top height).

However, there was no obvious way to stop it once it had taken off, which would be a great disadvantage in any non-disposable application. Also, because the reactor was unshielded, it was dangerous to be in or around the flight path of the vehicle (although the exhaust itself wasn't radioactive). These disadvantages limit the application to warhead delivery system for all-out nuclear war, which it was being designed for.

Scramjets

Scramjets are an evolution of ramjets that are able to operate at much higher speeds than any other kind of airbreathing engine. They share a similar structure with ramjets, being a specially shaped tube that compresses air with no moving parts through ram-air compression. Scramjets, however, operate with supersonic airflow through the entire engine. Thus, scramjets do not have the diffuser required by ramjets to slow the incoming airflow to subsonic speeds.

Scramjets start working at speeds of at least Mach 4, and have a maximum useful speed of approximately Mach 17. Due to aerodynamic heating at these high speeds, cooling poses a challenge to engineers.

Turborocket

The air turborocket is a form of combined-cycle jet engine. The basic layout includes a gas generator, which produces high pressure gas, that drives a turbine/compressor assembly which compresses atmospheric air into a combustion chamber. This mixture is then combusted before leaving the device through a nozzle and creating thrust. There are many different types of air turborockets. The various types generally differ in how the gas generator section of the engine functions.

Air turborockets are often referred to as turboramjets, turboramjet rockets, turborocket expanders, and many others. As there is no consensus on which names apply to which specific concepts, various sources may use the same name for two different concepts.

Terminology

To describe the RPM of a jet engine, abbreviations are commonly used:

- For a turboprop engine, N_p refers to the RPM of the propeller shaft. For example, a common N_p would be about 2200 RPM for a constant speed propeller.

- N_1 or N_g refers to the speed of the gas generator (gas producer) section (RPM). Each engine manufacturer will pick between those two abbreviation but N1 is mainly used for turbofan engines whereas Ng is mainly used for turboprop or turboshaft engines. For example, a common N_p would be on the order of 30,000 RPM.
- N_2 or N_f refers to the speed of the power turbine section. Each engine manufacturer will pick between those two abbreviations but N2 is mainly used for turbofan engine where Nf is mainly used for turboprop or turboshaft engines. In many cases, even for free turbine engines, the N_1 and N_2 may be very similar.
- N_s refers to the speed of the reduction gear box (RGB) output shaft for turboshaft engines.

In many cases, instead of expressing N-speeds (N_1, N_2) as a sheer RPM on cockpit displays, pilots are provided with the N-speeds expressed as a percentage of a nominal or maximal value. For example, at full power, the N_1 might be 101.5% or 100%.

This user interface decision has been made as a human factors consideration, since pilots are more likely to notice a problem with a two- or 3-digit percentage (where 100% implies a nominal value) than with a large, unbounded scalar number.

Turbojet

The turbojet is the oldest kind of general-purpose airbreathing jet engine. Two engineers, Frank Whittle in the United Kingdom and Hans von Ohain in Germany, developed the concept independently into practical engines during the late 1930s.

Turbojets consist of an air inlet, an air compressor, a combustion chamber, a gas turbine (that drives the air compressor) and a nozzle. The air is compressed into the chamber, heated and expanded by the fuel combustion and then allowed to expand out through the turbine into the nozzle where it is accelerated to high speed to provide propulsion.

Turbojets are quite inefficient if flown below about Mach 2 and very noisy. Most modern aircraft use turbofans instead for economic reasons. Turbojets are still very common in medium range cruise missiles, due to their high exhaust speed, low frontal area and relative simplicity.

History

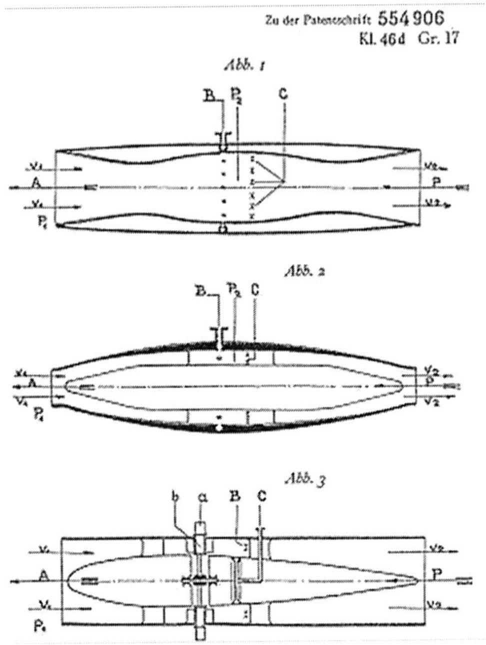

Figure 10: Albert Fono's German patent for jet Engines (January 1928). The third illustration is a turbojet

The first patent for using a gas turbine to power an aircraft was filed in 1921 by Frenchman Maxime Guillaume. His engine was to be an axial-flow turbojet, but was never constructed, as it would have required considerable advances over the state of the art in compressors.

Practical axial compressors were made possible by ideas from A.A.Griffith in a seminal paper in 1926 ("An Aerodynamic Theory of Turbine Design").

On 27 August 1939 the Heinkel He 178 became the world's first aircraft to fly under turbojet power with test-pilot Erich Warsitz at the controls, thus becoming the first practical jet plane. The first two operational turbojet aircraft, the Messerschmitt Me 262 and then the Gloster Meteor entered service towards the end of World War II in 1944.

A turbojet engine is used primarily to propel aircraft, but has been used for other vehicles, such as cars. Air is drawn into the rotating compressor via the intake and is compressed to a higher pressure before entering the combustion chamber. Fuel is mixed with the compressed air and ignited by a flame in the eddy of a flame holder. This combustion process significantly raises the temperature of the gas. Hot combustion products leaving the combustor expand through the turbine where power is extracted to drive the compressor. Although this expansion process reduces the turbine exit gas temperature and pressure, both parameters are usually still well above ambient conditions. The gas stream exiting the turbine expands to ambient pressure via the propelling nozzle, producing a high velocity jet in the exhaust plume. If the momentum of the exhaust stream exceeds the momentum of the intake stream, the impulse is positive, thus, there is a net forward thrust upon the airframe.

Early generation jet engines were pure turbojets, designed initially to use a centrifugal compressor (as in the Heinkel HeS 3), and very shortly afterwards began to use Axial compressors (as in the Junkers Jumo 004) for a smaller diametre to the overall engine housing. They were used because they were able to achieve very high altitudes and speeds, much higher than propeller engines, because of a better compression ratio and because of their high exhaust speed. However, they were not very fuel efficient. Modern jet engines are mainly turbofans, where a proportion of the air entering the intake bypasses the combustor; this proportion depends on the engine's bypass ratio. This makes turbofans much more efficient than turbojets at high subsonic/transonic and low supersonic speeds.

One of the most recent uses of turbojet engines was the Olympus 593 on Concorde. Concorde used turbojet engines because it turns out that the small cross-section and high exhaust speed is ideal for operation at Mach 2. Concorde's engine burnt less fuel to produce a given thrust for a mile at Mach 2.0 than a modern high-bypass turbofan such as General Electric CF6 at its Mach 0.86 optimum speed. Concorde's airframe, however, was far less efficient than that of any subsonic airliner.

Turbojet engines had a significant impact on commercial aviation. Aside from being faster than piston engines, turbojets had greater reliability, with some models demonstrating dispatch reliability rating in excess of 99.9%. Pre-jet commercial aircraft were designed with as many as 4 engines in part because of concerns over in-flight failures. Overseas flight paths were plotted to keep planes within an hour of a landing field, lengthening flights. Turbojets' reliability allowed for three and two-engine designs, and more direct long-distance flights.

Although ramjet engines are simpler in design as they have virtually no moving parts, they are incapable of operating at low flight speeds.

Early Designs

Early German engines had serious problems controlling the turbine inlet temperature. A lack of suitable alloys due to war shortages meant the turbine rotor and stator blades would sometimes disintegrate on first operation and never lasted long. Their early engines averaged 10–25 hours of operation before failing, often with chunks of metal flying out the back of the engine when the turbine overheated. British engines tended to fare better, running for 150 hours between overhauls. A few of the original fighters still exist with their original engines, but many have been re-engined with more modern engines with greater fuel efficiency and a longer TBO (such as the reproduction Me-262 powered by General Electric J85s).

Figure 10: J85-GE-17A turbojet engine from General Electric (1970)

The United States had the best materials because of their reliance on turbo/supercharging in high altitude bombers of World War II. For a time some US jet engines included the ability to inject water into the engine to cool the compressed flow before combustion, usually during takeoff. The water would tend to prevent complete combustion and as a result the engine ran cooler again, but the planes would take off leaving a huge plume of smoke.

Today these problems are much better handled, but temperature still limits turbojet airspeeds in supersonic flight. At the very highest speeds, the compression of the intake air raises the temperatures throughout the engine to the point that the turbine blades would melt, forcing a reduction in fuel flow to lower temperatures, but giving a reduced thrust and thus limiting the top speed. Ramjets and scramjets do not have turbine blades; therefore they are able to fly faster, and rocket engines run even hotter still. At lower speeds, better materials have increased the critical temperature, and automatic fuel management controls have made it nearly impossible to overheat the engine.

Design—Air Intake

Preceding the compressor is the air intake (or inlet). It is designed to be as efficient as possible at recovering the ram pressure of the air stream tube approaching the intake. The air leaving the intake then enters the compressor. The stators (stationary blades) guide the airflow of the compressed gases.

Compressor

The compressor is driven by the turbine. The compressor rotates at very high speed, adding energy to the airflow and at the same time squeezing (compressing) it into a smaller space. Compressing the air increases its pressure and temperature.

In most turbojet-powered aircraft, bleed air is extracted from the compressor section at various stages to perform a variety of jobs including air conditioning/pressurisation, engine inlet anti-icing and turbine cooling. Bleeding air off decreases the overall efficiency of the engine, but the usefulness of the compressed air outweighs the loss in efficiency. Several types of compressor are used in turbojets and gas turbines in general: axial, centrifugal, axial-centrifugal, double-centrifugal, etc.

Early turbojet compressors had overall pressure ratios as low as 5:1 (as do a lot of simple auxiliary power units and small propulsion turbojets today). Aerodynamic improvements, plus splitting the compression system into two separate units and/or incorporating variable compressor geometry, enabled later turbojets to have overall pressure ratios of 15:1 or more. For comparison, modern civil turbofan engines have overall pressure ratios of 44:1 or more.

After leaving the compressor section, the compressed air enters the combustion chamber.

Combustion Chamber

The burning process in the combustor is significantly different from that in a piston engine. In a piston engine the burning gases are confined to a small volume and, as the fuel burns, the pressure increases dramatically. In a turbojet the air and fuel mixture passes unconfined through the combustion chamber. As the mixture burns its temperature increases dramatically, but the pressure actually decreases a few percent.

The fuel-air mixture must be brought almost to a stop so that a stable flame can be maintained. This occurs just after the start of the combustion chamber. The aft part of this flame front is allowed to progress rearward. This ensures that all of the fuel is burned, as the flame becomes hotter when it leans out, and because of the shape of the combustion chamber the flow is accelerated rearwards. Some pressure drop is required, as it is the reason why the expanding gases travel out the rear of the engine rather than out the front. Less than 25% of the air is involved in combustion, in some engines as little as 12%, the rest acting as a reservoir to absorb the heating effects of the burning fuel.

Another difference between piston engines and jet engines is that the peak flame temperature in a piston engine is experienced only momentarily in a small portion of the full cycle. The combustor in a jet engine is exposed to the peak flame temperature continuously and operates at a pressure high enough that a stoichiometric fuel-air ratio would melt the can and everything downstream. Instead, jet engines run a very lean mixture, so lean that it would not normally support combustion. A central core of the flow (primary airflow) is mixed with enough fuel to burn readily. The cans are carefully shaped to maintain a layer of fresh unburned air between the metal surfaces and the central core. This unburned air (secondary airflow) mixes into the burned gases to bring the temperature down to something a turbine can tolerate.

Turbine

Hot gases leaving the combustor are allowed to expand through the turbine. Turbines are usually made up of high temperature metals such as inconel to resist the high temperature, and frequently have built-in cooling channels. In the first stage the turbine is largely an impulse turbine (similar to a pelton wheel) and rotates because of the impact of the hot gas stream. Later stages are convergent ducts that accelerate the gas rearward and gain energy from that process. Pressure drops, and energy is transferred into the shaft. The turbine's rotational energy is used primarily to drive the compressor. Some shaft power is extracted to drive accessories, like fuel, oil, and hydraulic pumps. Because of its significantly higher entry temperature, the turbine pressure ratio is much lower than that of the compressor. In a turbojet almost two-thirds of all the power generated by burning fuel is used by the compressor to compress the air for the engine.

Nozzle

After the turbine, the gases are allowed to expand through the exhaust nozzle to atmospheric pressure, producing a high velocity jet in the exhaust plume. In a convergent nozzle, the ducting narrows progressively to a throat. The nozzle pressure ratio on a turbojet is usually high enough for the expanding gases to reach Mach 1.0 and choke the throat. Normally, the flow will go supersonic in the exhaust plume outside the engine.

If, however, a convergent-divergent de Laval nozzle is fitted, the divergent (increasing flow area) section allows the gases to reach supersonic velocity within the nozzle itself. This is slightly more efficient on thrust than using a convergent nozzle. There is, however, the added weight and complexity since the convergent-divergent nozzle must be fully variable in its shape to cope with changes in gas flow caused by engine throttling.

Afterburner

An afterburner or "reheat jetpipe" is a device added to the rear of the jet engine. It provides a means of spraying fuel directly into the hot exhaust, where it ignites and boosts available thrust significantly; a drawback is its very high fuel consumption rate. Afterburners are used almost exclusively on supersonic aircraft – most of these are military aircraft. The two supersonic civilian transports, Concorde and the TU-144, also utilised afterburners but these two have now been retired from service. Scaled Composites White Knight, a carrier aircraft for the experimental SpaceShipOne suborbital spacecraft, also utilises an afterburner.

Thrust Reverser

A thrust reverser is, essentially, a pair of clamshell doors mounted at the rear of the engine which, when deployed, divert thrust normal to the jet engine flow to help slow an aircraft upon landing. They are often used in conjunction with spoilers. The accidental deployment of a thrust reverser during flight is a dangerous event that can lead to loss of control and destruction of the aircraft. Thrust reversers are more convenient than drogue parachutes, though mechanically more complex and expensive.

Net Thrust

The net thrust F_N of a turbojet is given by:

$$F_N = (\dot{m}_{air} + \dot{m}_f)V_j - \dot{m}_{air}V$$

where:

$\dot{m}_{air}$	is the rate of flow of air through the engine
$\dot{m}_f$	is the rate of flow of fuel entering the engine

V_j	is the speed of the jet (the exhaust plume) and is assumed to be less than sonic velocity
V	is the true airspeed of the aircraft
$(\dot{m}_{air} + \dot{m}_f)V_j$	represents the nozzle gross thrust
$\dot{m}_{air}V$	represents the ram drag of the intake

If the speed of the jet is equal to sonic velocity the nozzle is said to be choked. If the nozzle is *choked* the pressure at the nozzle exit plane is greater than atmospheric pressure, and extra terms must be added to the above equation to account for the *pressure thrust.*

The rate of flow of fuel entering the engine is very small compared with the rate of flow of air. If the contribution of fuel to the nozzle gross thrust is ignored, the net thrust is:

$$F_N = \dot{m}_{air}(V_j - V)$$

The speed of the jet V_j must exceed the true airspeed of the aircraft V if there is to be a net forward thrust on the airframe. The speed V_j can be calculated thermodynamically based on adiabatic expansion.

A simple turbojet engine will produce thrust of approximately: 2.5 pounds force per horsepower (15 mN/W).

Cycle Improvements

Thermodynamics of a typical air-breathing jet engine are modelled approximately by a Brayton Cycle.

Increasing the overall pressure ratio of the compression system raises the combustor entry temperature. Therefore, at a fixed fuel flow and airflow, there is an increase in turbine inlet temperature. Although the higher temperature rise across the compression system, implies a larger temperature drop over the turbine system, the nozzle temperature is unaffected, because the same amount of heat is being added to the system. There is, however, a rise in nozzle pressure, because overall pressure ratio increases faster than the turbine expansion ratio. Consequently, net thrust increases, while specific fuel consumption (fuel flow/net thrust) decreases.

Thus turbojets can be made more fuel efficient by raising overall pressure ratio and turbine inlet temperature in union. However, better turbine materials and/or improved vane/blade cooling are required to cope with increases in both turbine inlet temperature and compressor delivery temperature. Increasing the latter requires better compressor materials. Minimising heat losses and optimising the inlet temperature ratio will increase the system's useful work and the thermal efficiency of the turbo jet engine.

Turbofan

The turbofan is a type of airbreathing jet engine that is widely used for aircraft propulsion. A turbofan combines two types of engines, the *turbo* portion which is a conventional gas turbine engine, and the *fan*, a propeller-like ducted fan. The engine produces thrust through a combination of these two portions working in concert; engines that use more jet thrust relative to fan thrust are known as *low bypass turbofans*, while those that have considerably more fan thrust than jet are known as *high bypass*. Most commercial aviation jet engines in use today are of the high-bypass type, and most modern military engines are low-bypass,

The fan serves two duties. Part of the airstream from the fan passes through the core, providing oxygen to burn fuel to create power. However, the rest of the air flow bypasses the engine core and mixes with the faster stream from the core at the back of the engine. As engine noise is a function of exhaust temperature, turbofan engines are significantly quieter than a pure-jet of the same thrust. Additionally, the efficiency of propulsion is a function of the relative airspeed of the exhaust to the surrounding air; propellers are most efficient for low speed, pure jets for high speeds, and ducted fans in the middle. Turbofans are thus the most efficient engines in the range of speeds from about 500 to 1000 km/h, the speed at which most commercial aircraft operate. Turbofans retain an efficiency edge over pure jets at low supersonic speeds up to roughly Mach 1.6, but have also been found to be efficient when used with continuous afterburner at Mach 3 and above. However, the lower exhaust speed also reduces thrust at high vehicle speeds.

The vast majority of turbofans follow the same basic design with a large fan at the front of the engine with a relatively small jet engine behind it. There have been a number of variations on this theme, however, including rear-mounted fans where they can be easily added to an existing pure-jet design, or designs that combine a low-pressure turbine and a fan stage in a single rear-mounted unit.

Unlike a reciprocating engine, a turbojet uses continuous-flow combustion. In a single-spool (or single-shaft) turbojet, which is the most basic form and the earliest type of turbojet to be developed, air enters an intake before being compressed to a higher pressure by a rotating (fan-like) compressor. The compressed air passes on to a combustor, where it is mixed with a fuel (*e.g.* kerosene) and ignited. The hot combustion gases then enter a windmill-like turbine, where power is extracted to drive the compressor. Although the expansion process in the turbine reduces the gas pressure (and temperature) somewhat, the remaining energy and pressure are employed to provide a high-velocity jet by passing the gas through a propelling nozzle. This process produces a net thrust opposite to the direction of the jet flow.

After World War II, two-spool (or two-shaft) turbojets were developed to make it easier to throttle back compression systems with a high design overall pressure ratio (*i.e.,* combustor inlet pressure/intake delivery pressure). Adopting the two-spool arrangement enables the compression system to be split in two, with a low pressure (LP) compressor supercharging a high pressure (HP) compressor. Each compressor is mounted on a separate (co-axial) shaft, driven by its own turbine (i.e. the HP turbine and LP turbine). Otherwise, a two-spool turbojet is much like a single-spool engine.

Modern turbofans evolved from the two-spool axial-flow turbojet engine, essentially by increasing the relative size of the low pressure (LP) compressor to the point where some (if not most) of the air exiting the unit actually bypasses the core (or gas-generator) stream passing through the main combustor. This bypass air either expands through a separate propelling nozzle, or is mixed with the hot gases leaving the low pressure (LP) turbine, before expanding through a mixed stream propelling nozzle. Owing to a lower jet velocity, a modern civil turbofan is quieter than the equivalent turbojet. Turbofans also have a better thermal efficiency, which is explained later in the article. In a turbofan, the LP compressor is often called a fan. Civil-aviation turbofans usually have a single fan stage, whereas most military-aviation turbofans (e.g. combat and trainer aircraft applications) have multistage fans. Modern military transport turbofan engines are similar to those that propel civil jetliners.

Turboprop engines are gas-turbine engines that deliver almost all of their power to a shaft to drive a propeller. Turboprops remain popular on very small or slow aircraft, such as small commuter airliners, for their fuel efficiency at lower speeds, as well as on medium military transports and patrol planes, such as the C-130 Hercules and P-3 Orion, for their high takeoff performance and mission endurance benefits.

If the turboprop is better at moderate flight speeds and the turbojet is better at very high speeds, it might be imagined that at some speed range in the middle a mixture of the two is best. Such an engine is the turbofan. Another name sometimes used is ducted fan, though that term is also used for propellers and fans used in vertical-flight applications.

The difference between a turbofan and a propeller, besides direct thrust, is that the intake duct of the former slows the air before it arrives at the fan face. As both propeller and fan blades must operate at subsonic inlet velocities to be efficient, ducted fans allow efficient operation at higher vehicle speeds.

Figure 11: Duct work on an Dassault/Dornier Alpha Jet — the increasing diametre of the inlet duct slows incoming air according to the principle of continuity. As the incoming air slows, its static pressure increases according to Bernoulli's Principle.

Depending on specific thrust, ducted fans operate best from about 400 to 2,000 km/h (250 to 1,300 mph), which is why turbofans are the most common type of engine for aviation use today—in airliners as well as in subsonic and supersonic military fighter and trainer aircraft. It should be noted, however, that turbofans use extensive ducting to force incoming air to subsonic velocities (thus reducing shock waves throughout the engine).

Bypass ratio (bypassed airflow to combustor airflow) is a parameter often used for classifying turbofans; when the low-bypass Conway engine entered service in 1960, no one even called it a turbofan, that term first being applied to Pratt and Whitney's JT3D with its 1-to-1 bypass.

The noise of any type of jet engine is strongly related to the velocity of the exhaust gases, typically being proportional to the eighth power of the jet velocity. High-bypass-ratio (*i.e.,* low-specific-thrust) turbofans are relatively quiet compared to turbojets and low-bypass-ratio (*i.e.,* high-specific-thrust) turbofans. A low-specific-thrust engine has a low jet velocity by definition, as the following approximate equation for net thrust implies:

$$F_n = \dot{m}\,(V_{jfe} - V_a)$$

where: $\dot{m}$

= intake mass flow V_{jfe}

= fully expanded jet velocity (in the exhaust plume) V_a

= aircraft flight velocity

Rearranging the above equation, specific thrust is given by: $\frac{F_n}{\dot{m}} = (V_{jfe} - V_a)$

So for zero flight velocity, specific thrust is directly proportional to jet velocity. Relatively speaking, low-specific-thrust engines are large in diametre to accommodate the high airflow required for a given thrust.

Although jet aircraft are loud, a conventional piston engine or a turboprop engine delivering the same thrust would be much louder.

Early Turbofans

Early turbojet engines were very fuel-inefficient, as their overall pressure ratio and turbine inlet temperature were severely limited by the technology available at the time. The very first running turbofan was the German Daimler-Benz DB 670 (designated as the 109-007 by the RLM) which was operated on its testbed on April 1, 1943. The engine was abandoned later while the war went on and problems could not be solved. The British wartime Metrovick F.2 axial flow jet was given a fan to create the first British turbofan.

Improved materials, and the introduction of twin compressors such as in the Pratt & Whitney JT3C engine, increased the overall pressure ratio and thus the thermodynamic efficiency of engines, but they also led to a poor propulsive efficiency, as pure turbojets have a high specific thrust/high velocity exhaust better suited to supersonic flight.

The original low-bypass turbofan engines were designed to improve propulsive efficiency by reducing the exhaust velocity to a value closer to that of the aircraft. The Rolls-Royce Conway, the first production turbofan, had a bypass ratio of 0.3, similar to the modern General Electric F404 fighter engine. Civilian turbofan engines of the 1960s, such as the Pratt & Whitney JT8D and the Rolls-Royce Spey had bypass ratios closer to 1, but were not dissimilar to their military equivalents.

The unusual General Electric CF700 turbofan engine was developed as an aft-fan engine with a 2.0 bypass ratio. This was derived from the General Electric J85/CJ610 turbojet (2,850 lbf or 12,650 N) to power the larger Rockwell Sabreliner 75/80 model aircraft, as well as the Dassault Falcon 20 with about a 50% increase in thrust (4,200 lbf or 18,700 N). The CF700 was the first small turbofan in the world to be certified by the Federal Aviation Administration (FAA). There are now over 400 CF700 aircraft in operation around the world, with an experience base of over 10 million service hours. The CF700 turbofan engine was also used to train Moon-bound astronauts in Project Apollo as the powerplant for the Lunar Landing Research Vehicle.

Low-bypass Turbofan

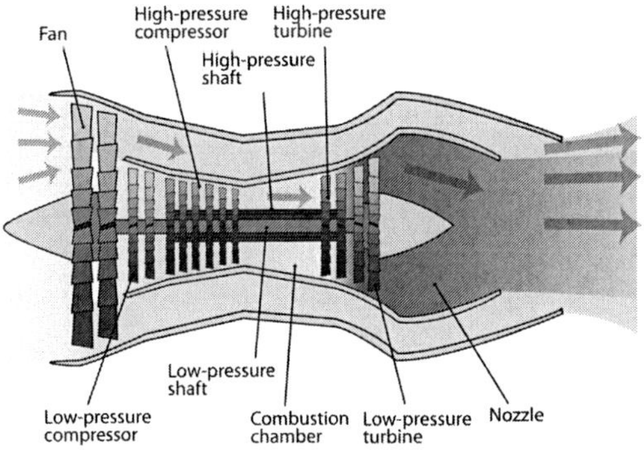

Figure 12: Schematic diagram illustrating a 2-spool, low-bypass turbofan engine with a mixed exhaust, showing the low-pressure (green) and high-pressure (purple) spools. The fan (and booster stages) are driven by the low-pressure turbine, whereas the high-pressure compressor is powered by the high-pressure turbine

A high specific thrust/low bypass ratio turbofan normally has a multi-stage fan, developing a relatively high pressure ratio and, thus, yielding a high (mixed or cold) exhaust velocity. The core airflow needs to be large enough to give sufficient core power to drive the fan. A smaller core flow/higher bypass ratio cycle can be achieved by raising the (HP) turbine rotor inlet temperature.

Imagine a retrofit situation where a new low bypass ratio, mixed exhaust, turbofan is replacing an old turbojet, in a particular military application. Say the new engine is to have the same airflow and net thrust (i.e. same specific thrust) as the one it is replacing. A bypass flow can only be introduced if the turbine inlet temperature is allowed to increase, to compensate for a correspondingly smaller core flow. Improvements in turbine cooling/material technology would facilitate the use of a higher turbine inlet temperature, despite increases in cooling air temperature, resulting from a probable increase in overall pressure ratio.

Efficiently done, the resulting turbofan would probably operate at a higher nozzle pressure ratio than the turbojet, but with a lower exhaust temperature to retain net thrust. Since the temperature rise across the whole engine (intake to nozzle) would be lower, the (dry power) fuel flow would also be reduced, resulting in a better specific fuel consumption (SFC).

A few low-bypass ratio military turbofans (e.g. F404) have Variable Inlet Guide Vanes, with piano-style hinges, to direct air onto the first rotor stage. This improves the fan surge margin in the mid-flow range. The swing wing F-111 achieved a very high range/payload capability by pioneering this, and it was also the heart of the famous F-14 Tomcat air superiority fighter which used the same engines in a smaller, more agile airframe to achieve efficient cruise and Mach 2 speed.

Afterburning Turbofan

Since the 1970s, most jet fighter engines have been low/medium bypass turbofans with a mixed exhaust, afterburner and variable area final nozzle. An afterburner is a combustor located downstream of the turbine blades and directly upstream of the nozzle, which burns fuel from afterburner-specific fuel injectors. When lit, prodigious amounts of fuel are burnt in the afterburner, raising the temperature of exhaust gases by a significant degree, resulting in a higher exhaust velocity/ engine specific thrust. The variable geometry nozzle must open to a larger throat area to accommodate the extra volume flow when the afterburner is lit. Afterburning is often designed to give a significant thrust boost for take off, transonic acceleration and combat maneuvers, but is very fuel intensive. Consequently afterburning can only be used for short portions of a mission.

Unlike the main combustor, where the downstream turbine blades must not be damaged by high temperatures, an afterburner can operate at the ideal maximum (stoichiometric) temperature. At a fixed total applied fuel: air ratio, the total fuel flow for a given fan airflow will be the same, regardless of the dry specific thrust of the engine. However, a high specific thrust turbofan will, by definition, have a higher nozzle pressure ratio, resulting in a higher afterburning net thrust and, therefore, a lower afterburning specific fuel consumption. However, high specific thrust engines have a high dry SFC.

The situation is reversed for a medium specific thrust afterburning turbofan: i.e. poor afterburning SFC/good dry SFC. The former engine is suitable for a combat aircraft which must remain in afterburning combat for a fairly long period, but only has to fight fairly close to the airfield (e.g. cross border skirmishes) The latter engine is better for an aircraft that has to fly some distance, or loiter for a long time, before going into combat. However, the pilot can only afford to stay in afterburning for a short period, before aircraft fuel reserves become dangerously low.

Modern low-bypass military turbofans include the Pratt & Whitney F119, the Eurojet EJ200, the General Electric F110, the Klimov RD-33, and the Saturn AL-31, all of which feature a mixed exhaust, afterburner and variable area propelling nozzle.

High-bypass Turbofan

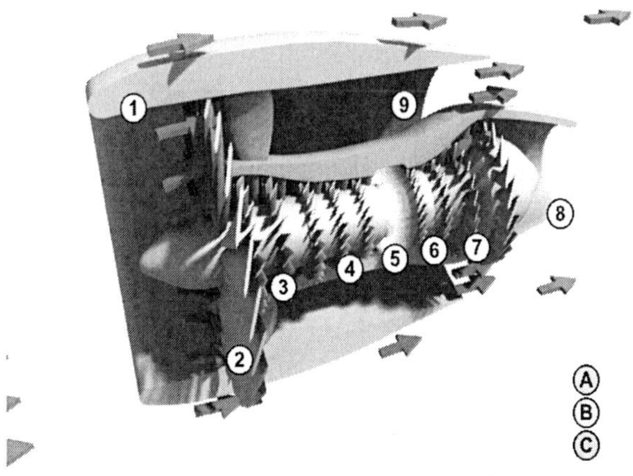

Figure 13: Animation of a 2-spool, high-bypass turbofan.

A. Low pressure spool

B. High pressure spool

C. Stationary components
   1. Nacelle
   2. Fan
   3. Low pressure compressor
   4. High pressure compressor
   5. Combustion chamber
   6. High pressure turbine
   7. Low pressure turbine
   8. Core nozzle
   9. Fan nozzle.

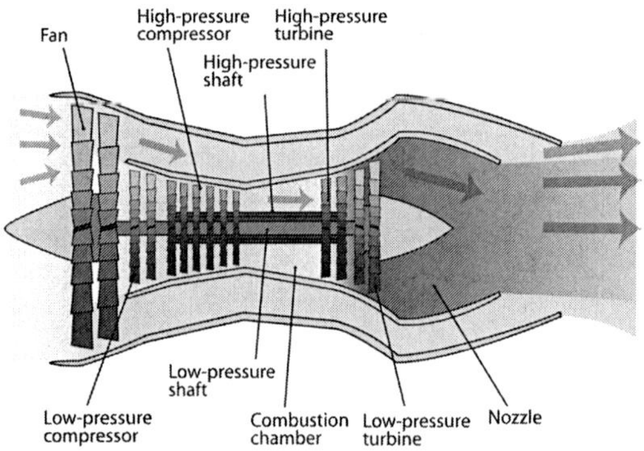

Figure 14: Schematic diagram illustrating a 2-spool, low-bypass turbofan engine with a mixed exhaust, showing the low-pressure (green) and high-pressure (purple) spools. The fan (and booster stages) are driven by the low-pressure turbine, whereas the high-pressure compressor is powered by the high-pressure turbine

The low specific thrust/high bypass ratio turbofans used in today's civil jetliners (and some military transport aircraft) evolved from the high specific thrust/low bypass ratio turbofans used in such [production] aircraft back in the 1960s.

Low specific thrust is achieved by replacing the multi-stage fan with a single stage unit. Unlike some military engines, modern civil turbofans do not have any stationary inlet guide vanes in front of the fan rotor. The fan is scaled to achieve the desired net thrust.

The core (or gas generator) of the engine must generate sufficient core power to at least drive the fan at its design flow and pressure ratio. Through improvements in turbine cooling/material technology, a higher (HP) turbine rotor inlet temperature can be used, thus facilitating a smaller (and lighter) core and (potentially) improving the core thermal efficiency. Reducing the core mass flow tends to increase the load on the LP turbine, so this unit may require additional stages to reduce the average stage loading and to maintain LP turbine efficiency. Reducing core flow also increases bypass ratio (5:1, or more, is now common).

Further improvements in core thermal efficiency can be achieved by raising the overall pressure ratio of the core. Improved blade aerodynamics reduces the number of extra compressor stages required. With multiple compressors (i.e. LPC, IPC, HPC) dramatic increases in overall pressure ratio have become possible. Variable geometry (i.e. stators) enable high pressure ratio compressors to work surge-free at all throttle settings.

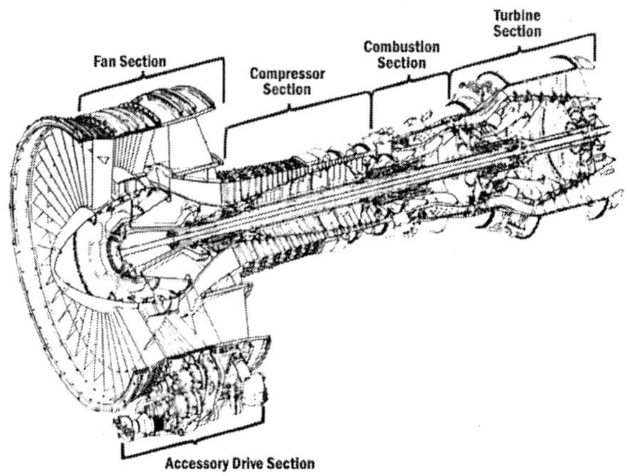

Figure 15: Cutaway diagram of the General Electric CF6-6 engine

The first high-bypass turbofan engine was the General Electric TF39, designed in mid 1960s to power the Lockheed C-5 Galaxy military transport aircraft. The civil General Electric CF6 engine used a derived design. Other high-bypass turbofans are the Pratt & Whitney JT9D, the three-shaft Rolls-Royce RB211 and the CFM International CFM56. More recent large high-bypass turbofans include the Pratt & Whitney PW4000, the three-shaft Rolls-Royce Trent, the General Electric GE90/GEnx and the GP7000, produced jointly by GE and P&W.

High-bypass turbofan engines are generally quieter than the earlier low bypass ratio civil engines. This is not so much due to the higher bypass ratio as to the use of a low pressure ratio, single stage fan which significantly reduces specific thrust and, thereby, jet velocity. The combination of a higher overall pressure ratio and turbine inlet temperature improves thermal efficiency. This, together with a lower specific thrust (better propulsive efficiency), leads to a lower specific fuel consumption.

For reasons of fuel economy, and also of reduced noise, almost all of today's jet airliners are powered by high-bypass turbofans. Although modern combat aircraft tend to use low bypass ratio turbofans, military transport aircraft (e.g. C-17) mainly use high bypass ratio turbofans (or turboprops) for fuel efficiency.

Because of the implied low mean jet velocity, a high bypass ratio/ low specific thrust turbofan has a high thrust lapse rate (with rising flight speed). Consequently the engine must be over-sized to give sufficient thrust during climb/cruise at high flight speeds (e.g. Mach 0.83). Because of the high thrust lapse rate, the static (i.e. Mach 0) thrust is relatively high. This enables heavily laden, wide body aircraft to accelerate quickly during take-off and consequently lift-off within a reasonable runway length.

The turbofans on twin engined airliners are further over-sized to cope with losing one engine during take-off, which reduces the aircraft's net thrust by 50%. Modern twin engined airliners normally climb very steeply immediately after take-off. If one engine is lost, the climb-out is much shallower, but sufficient to clear obstacles in the flightpath.

The Soviet Union's engine technology was less advanced than the West's and its first wide-body aircraft, the Ilyushin Il-86, was powered by low-bypass engines. The Yakovlev Yak-42, a medium-range, rear-engined aircraft seating up to 120 passengers introduced in 1980 was the first Soviet aircraft to use high-bypass engines.

Turbofan Configurations

Turbofan engines come in a variety of engine configurations. For a given engine cycle (i.e. same airflow, bypass ratio, fan pressure ratio, overall pressure ratio and HP turbine rotor inlet temperature), the choice of turbofan configuration has little impact upon the design point performance (e.g. net thrust, SFC), as long as overall component performance is maintained. Off-design performance and stability is, however, affected by engine configuration.

As the design overall pressure ratio of an engine cycle increases, it becomes more difficult to throttle the compression system, without encountering an instability known as compressor surge. This occurs when some of the compressor aerofoils stall (like the wings of an aircraft) causing a violent change in the direction of the airflow. However, compressor stall can be avoided, at throttled conditions, by progressively:

1. opening interstage/intercompressor blow-off valves (inefficient) and/or
2. closing variable stators within the compressor.

Most modern American civil turbofans employ a relatively high pressure ratio High Pressure (HP) Compressor, with many rows of variable stators to control surge margin at part-throttle. In the three-spool RB211/Trent the core compression system is split into two, with the IP compressor, which supercharges the HP compressor, being on a different coaxial shaft and driven by a separate (IP) turbine.

As the HP Compressor has a modest pressure ratio it can be throttled-back surge-free, without employing variable geometry. However, because a shallow IP compressor working line is inevitable, the IPC has one stage of variable geometry on all variants except the -535, which has none.

Single Shaft Turbofan

Although far from common, the Single Shaft Turbofan is probably the simplest configuration, comprising a fan and high pressure compressor driven by a single turbine unit, all on the same shaft. The SNECMA M53, which powers Mirage fighter aircraft, is an example of a Single Shaft Turbofan. Despite the simplicity of the turbomachinery configuration, the M53 requires a variable area mixer to facilitate part-throttle operation.

Aft-fan Turbofan

One of the earliest turbofans was a derivative of the General Electric J79 turbojet, known as the CJ805-23, which featured an integrated aft fan/low pressure (LP) turbine unit located in the turbojet exhaust jetpipe. Hot gas from the turbojet turbine exhaust expanded through the LP turbine, the fan blades being a radial extension of the turbine blades. This aft-fan configuration was later exploited in the General Electric GE-36 UDF (propfan) Demonstrator of the early 80's. One of the problems with the aft fan configuration is hot gas leakage from the LP turbine to the fan.

Basic Two Spool

Many turbofans have the Basic Two Spool configuration where both the fan and LP turbine (i.e. LP spool) are mounted on a second (LP) shaft, running concentrically with the HP spool (i.e. HP compressor driven by HP turbine). The BR710 is typical of this configuration. At the smaller thrust sizes, instead of all-axial blading, the HP compressor configuration may be axial-centrifugal, double-centrifugal or even diagonal/centrifugal.

Boosted Two Spool

Higher overall pressure ratios can be achieved by either raising the HP compressor pressure ratio or adding an Intermediate Pressure (IP) Compressor between the fan and HP compressor, to supercharge or boost the latter unit helping to raise the overall pressure ratio of the engine cycle to the very high levels employed today (i.e. greater than 40:1, typically). All of the large American turbofans feature an IP compressor mounted on the LP shaft and driven, like the fan, by the LP turbine, the mechanical speed of which is dictated by the tip speed and diametre of the fan. The Rolls Royce BR715 is a non american example of this. The high bypass ratios (i.e. fan duct flow/ core flow) used in modern civil turbofans tends to reduce the relative diametre of the attached IP compressor, causing its mean tip speed to decrease. Consequently more IPC stages are required to develop the necessary IPC pressure rise.

Three Spool

Rolls-Royce chose a three spool configuration for their large civil turbofans (i.e. the RB211 and Trent families), where the Intermediate Pressure (IP) compressor is mounted on a separate (IP) shaft, running concentrically with the LP and HP shafts, and is driven by a separate IP turbine.

Ivchenko Design Bureau chose the same configuration for their Lotarev D-36 engine, followed by Lotarev/Progress D-18T and Progress D-436. The Turbo-Union RB199 military turbofan also has a three spool configuration, as does the Russian military Kuznetsov NK-321.

Geared Fan

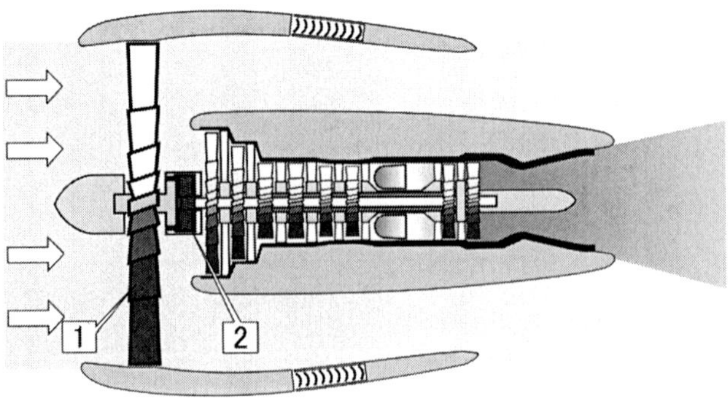

Figure 16: Geared turbofan

As bypass ratio increases, the mean radius ratio of the fan and LP turbine increases. Consequently, if the fan is to rotate at its optimum blade speed the LP turbine blading will spin slowly, so additional LPT stages will be required, to extract sufficient energy to drive the fan. Introducing a (planetary) reduction gearbox, with a suitable gear ratio, between the LP shaft and the fan enables both the fan and LP turbine to operate at their optimum speeds. Typical of this configuration are the long-established Honeywell TFE731, the Honeywell ALF 502/507, and the recent Pratt & Whitney PW1000G.

Military Turbofans

Most of the configurations discussed above are used in civil turbofans, while modern military turbofans (e.g. SNECMA M88) are usually Basic Two Spool.

High Pressure Turbine

Most civil turbofans use a high efficiency, 2-stage HP turbine to drive the HP compressor. The CFM56 uses an alternative approach: a single stage, high-work unit. While this approach is probably less efficient, there are savings on cooling air, weight and cost. In the RB211 and Trent series, Rolls-Royce split the two stages into two discrete units; one on the HP shaft driving the HP compressor; the other on the IP shaft driving the IP (Intermediate Pressure) Compressor. Modern military turbofans tend to use single stage HP turbines.

Low Pressure Turbine

Modern civil turbofans have multi-stage LP turbines (e.g. 3, 4, 5, 6, 7). The number of stages required depends on the engine cycle bypass ratio and how much supercharging (i.e. IP compression) is on the LP shaft, behind the fan. A geared fan may reduce the number of required LPT stages in some applications. Because of the much lower bypass ratios employed, military turbofans only require one or two LP turbine stages.

Cycle Improvements

Consider a mixed turbofan with a fixed bypass ratio and airflow. Increasing the overall pressure ratio of the compression system raises the combustor entry temperature. Therefore, at a fixed fuel flow there is an increase in (HP) turbine rotor inlet temperature. Although the higher temperature rise across the compression system implies a larger temperature drop over the turbine system, the mixed nozzle temperature is unaffected, because the same amount of heat is being added to the system. There is, however, a rise in nozzle pressure, because overall pressure ratio increases faster than the turbine expansion ratio, causing an increase in the hot mixer entry pressure. Consequently, net thrust increases, whilst specific fuel consumption (fuel flow/net thrust) decreases. A similar trend occurs with unmixed turbofans.

So turbofans can be made more fuel efficient by raising overall pressure ratio and turbine rotor inlet temperature in unison. However, better turbine materials and/or improved vane/blade cooling are required to cope with increases in both turbine rotor inlet temperature and compressor delivery temperature. Increasing the latter may require better compressor materials. Overall pressure ratio can be increased by improving fan (or) LP compressor pressure ratio and/or HP compressor pressure ratio. If the latter is held constant, the increase in (HP) compressor delivery temperature (from raising overall pressure ratio) implies an increase in HP mechanical speed. However, stressing considerations might limit this parameter, implying, despite an increase in overall pressure ratio, a reduction in HP compressor pressure ratio.

According to simple theory, if the ratio turbine rotor inlet temperature/(HP) compressor delivery temperature is maintained, the HP turbine throat area can be retained. However, this assumes that cycle improvements are obtained, whilst retaining the datum (HP) compressor exit flow function (non-dimensional flow). In practise, changes to the non-dimensional speed of the (HP) compressor and cooling bleed extraction would probably make this assumption invalid, making some adjustment to HP turbine throat area unavoidable. This means the HP turbine nozzle guide vanes would have to be different from the original! In all probability, the downstream LP turbine nozzle guide vanes would have to be changed anyway.

Thrust Growth

Thrust growth is obtained by increasing core power. There are two basic routes available:

a) hot route: increase HP turbine rotor inlet temperature

b) cold route: increase core mass flow.

Both routes require an increase in the combustor fuel flow and, therefore, the heat energy added to the core stream.

The hot route may require changes in turbine blade/vane materials and/or better blade/vane cooling. The cold route can be obtained by one of the following:

1. adding T-stages to the LP/IP compression
2. adding a zero-stage to the HP compression
3. improving the compression process, without adding stages (e.g. higher fan hub pressure ratio).

Alternatively, the core size can be increased, to raise core airflow, without changing overall pressure ratio. This route is expensive, since a new (upflowed) turbine system (and possibly a larger IP compressor) is also required.

Changes must also be made to the fan to absorb the extra core power. On a civil engine, jet noise considerations mean that any significant increase in Take-off thrust must be accompanied by a corresponding increase in fan mass flow, usually by increasing fan diametre. On military engines, the fan pressure ratio would probably be increased to improve specific thrust, jet noise not normally being an important factor.

Technical Discussion

1. Specific Thrust (net thrust/intake airflow) is an important parameter for turbofans and jet engines in general. Imagine a fan (driven by an appropriately sized electric motor) operating within a pipe, which is connected to a propelling nozzle. It is fairly obvious, the higher the Fan Pressure Ratio (fan discharge pressure/fan inlet pressure), the higher the jet velocity and the corresponding specific thrust. Now imagine we replace this set-up with an equivalent turbofan - same airflow and same fan pressure ratio. Obviously, the core of the turbofan must produce sufficient power to drive the fan via the Low Pressure (LP) Turbine. If we choose a low (HP) Turbine Inlet Temperature for the gas generator, the core airflow needs to be relatively high to compensate. The corresponding bypass ratio is therefore relatively low. If we raise the Turbine Inlet Temperature, the core airflow can be smaller, thus increasing bypass ratio. Raising turbine inlet temperature tends to increase thermal efficiency and, therefore, improve fuel efficiency.
2. Naturally, as altitude increases there is a decrease in air density and, therefore, the net thrust of an engine. There is also a flight speed effect, termed Thrust Lapse Rate. Consider the approximate equation for net thrust again:

$$F_n = m\,(V_{jfe} - V_a)$$

With a high specific thrust (e.g. fighter) engine, the jet velocity is relatively high, so intuitively one can see that increases in flight velocity have less of an impact upon net thrust than a medium specific thrust (e.g. trainer) engine, where the jet velocity is lower. The impact of thrust lapse rate upon a low specific thrust (e.g. civil) engine is even more severe. At high flight speeds, high specific thrust engines can pick-up net thrust through the ram rise in the intake, but this effect tends to diminish at supersonic speeds because of shock wave losses.

1. Thrust growth on civil turbofans is usually obtained by increasing fan airflow, thus preventing the jet noise becoming too high. However, the larger fan airflow requires more power from the core. This can be achieved by raising the Overall Pressure Ratio (combustor inlet pressure/intake delivery pressure) to induce more airflow into the core and by increasing turbine inlet temperature. Together, these parameters tend to increase core thermal efficiency and improve fuel efficiency.
2. Some high bypass ratio civil turbofans use an extremely low area ratio (less than 1.01), convergent-divergent, nozzle on the bypass (or mixed exhaust) stream, to control the fan working line. The nozzle acts as if it has variable geometry. At low flight speeds the nozzle is unchoked (less than a Mach Number of unity), so the exhaust gas speeds up as it approaches the throat and then slows down slightly as it reaches the divergent section. Consequently, the nozzle exit area controls the fan match and, being larger than the throat, pulls the fan working line slightly away from surge. At higher flight speeds, the ram rise in the intake increases nozzle pressure ratio to the point where the throat becomes choked (M=1.0). Under these circumstances, the throat area dictates the fan match and, being smaller than the exit, pushes the fan working line slightly towards surge. This is not a problem, since fan surge margin is much better at high flight speeds.
3. The off-design behaviour of turbofans is illustrated under compressor map and turbine map.
4. Because modern civil turbofans operate at low specific thrust, they only require a single fan stage to develop the required fan pressure ratio. The desired overall pressure ratio for the engine cycle is usually achieved by multiple axial stages on the core compression. Rolls-Royce tend to split the core compression

into two with an intermediate pressure (IP) supercharging the HP compressor, both units being driven by turbines with a single stage, mounted on separate shafts. Consequently, the HP compressor need only develop a modest pressure ratio (e.g.~4.5:1). US civil engines use much higher HP compressor pressure ratios (e.g. ~23:1 on the General Electric GE90) and tend to be driven by a two stage HP turbine. Even so, there are usually a few IP axial stages mounted on the LP shaft, behind the fan, to further supercharge the core compression system. Civil engines have multi-stage LP turbines, the number of stages being determined by the bypass ratio, the amount of IP compression on the LP shaft and the LP turbine blade speed.

5. Because military engines usually have to be able to fly very fast at Sea Level, the limit on HP compressor delivery temperature is reached at a fairly modest design overall pressure ratio, compared with that of a civil engine. Also the fan pressure ratio is relatively high, to achieve a medium to high specific thrust. Consequently, modern military turbofans usually only have 5 or 6 HP compressor stages and only require a single stage HP turbine. Low bypass ratio military turbofans usually have one LP turbine stage, but higher bypass ratio engines need two stages. In theory, by adding IP compressor stages, a modern military turbofan HP compressor could be used in a civil turbofan derivative, but the core would tend to be too small for high thrust applications.

Recent Developments in Blade Technology

The turbine blades in a turbofan engine are subject to high heat and stress, and require special fabrication. New material construction methods and material science have allowed blades, which were originally polycrystalline (regular metal), to be made from lined up metallic crystals and more recently mono-crystalline (i.e. single crystal) blades, which can operate at higher temperatures with less distortion.

Nickel-based superalloys are used for HP turbine blades in almost all modern jet engines. The temperature capabilities of turbine blades have increased mainly through four approaches: the manufacturing (casting) process, cooling path design, thermal barrier coating (TBC), and alloy development. Although turbine blade (and vane) materials have improved over the years, much of the increase in (HP) turbine inlet temperatures is due to improvements in blade/vane cooling technology. Relatively cool air is bled from the compression system, bypassing the combustion process, and enters the hollow blade or vane. After picking up heat from the blade/vane, the cooling air is dumped into the main gas stream. If the local gas temperatures are low enough, downstream blades/vanes are uncooled and not adversely affected.

Strictly speaking, cycle-wise the HP Turbine Rotor Inlet Temperature (after the temperature drop across the HPT stator) is more important than the (HP) turbine inlet temperature. Although some modern military and civil engines have peak RITs of the order of 1,560 °C (2,840 °F), such temperatures are only experienced for a short time (during take-off) on civil engines.

Turbofan Engine Manufacturers

The turbofan engine market is dominated by General Electric, Rolls-Royce plc and Pratt & Whitney, in order of market share. GE and SNECMA of France have a joint venture, CFM International which, as the 3rd largest manufacturer in terms of market share, fits between Rolls-Royce and Pratt & Whitney. Rolls-Royce and Pratt & Whitney also have a joint venture, International Aero Engines, specialising in engines for the Airbus A320 family, whilst finally, Pratt & Whitney and General Electric have a joint venture, Engine Alliance marketing a range of engines for aircraft such as the Airbus A380.

General Electric

GE Aviation, part of the General Electric Conglomerate, currently has the largest share of the turbofan engine market. Some of their engine models include the CF6, GE90 (only the Boeing 777) and GEnx engines. On the military side, GE engines power many U.S. military aircraft, including the F110, powering 80% of the US Air Force's F-16 Fighting Falcons, and the F404 and F414 engines, which power the Navy's F/A-18 Hornet and Super Hornet. Rolls-Royce and General Electric are jointly developing the F136 engine to power the Joint Strike Fighter.

CFM International

CFM International is a joint venture between GE Aircraft Engines and SNECMA of France. They have created the very successful CFM56 series, used on Boeing 737, Airbus A340, and Airbus A320 family aircraft.

Rolls-Royce

Rolls-Royce plc is the second largest manufacturer of turbofans and is most noted for their RB211 and Trent series, as well as their joint venture engines for the Airbus A320 and McDonnell Douglas MD-90 families (IAE V2500 with Pratt & Whitney and others), the Panavia Tornado (Turbo-Union RB199) and the Boeing 717 (BR700). The Rolls-Royce AE 3007 was developed by Allison Engine Company before its acquisition by Rolls-Royce, powers several Embraer regional jets. Rolls-Royce Trent 970s were the first engines to power the new Airbus A380. The famous thrust vectoring Pegasus engine is the primary powerplant of the Harrier "Jump Jet" and its derivatives.

Pratt & Whitney

Pratt & Whitney is third behind GE and Rolls-Royce in market share. The JT9D has the distinction of being chosen by Boeing to power the original Boeing 747 "Jumbo jet". The PW4000 series is the successor to the JT9D, and powers some Airbus A310, Airbus A300, Boeing 747, Boeing 767, Boeing 777, Airbus A330 and MD-11 aircraft. The PW4000 is certified for 180-minute ETOPS when used in twinjets. The first family has a 94-inch (2.4 m) fan diametre and is designed to power the Boeing 767, Boeing 747, MD-11, and the Airbus A300. The second family is the 100 inch (2.5 m) fan engine developed specifically for the Airbus A330 twinjet, and the third family has a diametre of 112-inch (2.8 m) designed to power Boeing 777. The Pratt & Whitney F119 and its derivative, the F135, power the United States Air Force's F-22 Raptor and the international F-35 Lightning II, respectively. Rolls-Royce are responsible for the lift fan which will provide the F-35B variants with a STOVL capability. The F100 engine was first used on the F-15 Eagle and F-16 Fighting Falcon. Newer Eagles and Falcons also come with GE F110 as an option, and the two are in competition.

Aviadvigatel

Aviadvigatel is a Russian manufacturer of aircraft engines that succeeded the Soviet Soloviev Design Bureau. The company currently offers several versions of the Aviadvigatel PS-90 engine that powers Ilyushin Il-96-300/400/400T, Tupolev Tu-204, Tu-214 series and the Ilyushin Il-76-MD-90. The company is also developing the new Aviadvigatel PD-14 engine for the new Russian MS-21 airliner.

Ivchenko-Progress

Ivchenko-Progress is the Ukrainian aircraft engine company that succeeded the Soviet Ivchenko Design Bureau. Some of their engine models include Progress D-436 available on the Antonov An-72/74, Yakovlev Yak-42, Beriev Be-200, Antonov An-148 and Tupolev Tu-334 and Progress D-18T that powers two of the world largest aeroplanes, Antonov An-124 and Antonov An-225.

Extreme Bypass Jet Engines

In the 1970s Rolls-Royce/SNECMA tested a M45SD-02 turbofan fitted with variable pitch fan blades to improve handling at ultra low fan pressure ratios and to provide thrust reverse down to zero aircraft speed. The engine was aimed at ultra quiet STOL aircraft operating from city centre airports.

In a bid for increased efficiency with speed, a development of the *turbofan* and *turboprop* known as a propfan engine was created that had an unducted fan.

The fan blades are situated outside of the duct, so that it appears like a turboprop with wide scimitar-like blades. Both General Electric and Pratt & Whitney/Allison demonstrated propfan engines in the 1980s. Excessive cabin noise and relatively cheap jet fuel prevented the engines being put into service.

Terminology

Afterburner : extra combustor immediately upstream of final nozzle (also called reheat).

Average stage loading : constant * (delta temperature)/[(blade speed) * (blade speed) * (number of stages)].

Bypass : airstream that completely bypasses the core compression system, combustor and turbine system.

Bypass ratio : bypass airflow /core compression inlet airflow.

Core : turbomachinery handling the airstream that passes through the combustor.

Core power : residual shaft power from ideal turbine expansion to ambient pressure after deducting core compression power.

Core thermal efficiency : core power/power equivalent of fuel flow.

Dry : afterburner (if fitted) not lit.

EGT : Exhaust Gas Temperature.

EPR : Engine Pressure Ratio.

Fan : turbofan LP compressor.

Fan pressure ratio : fan outlet total pressure/intake delivery total pressure.

Flex temp : use of artificially high apparent air temperature to reduce engine wear.

Gas generator : engine core.

HPC : high pressure compressor.

HP compressor : high pressure compressor.

HPT : high pressure turbine.

HP turbine : high pressure turbine.

Intake ram drag : penalty associated with jet engines picking up air from the atmosphere (conventional rocket motors do not have this drag term, because the oxidiser travels with the vehicle).

IEPR : Integrated Engine Pressure Ratio.

IPC : intermediate pressure compressor.

IP compressor : intermediate pressure compressor.

IPT : intermediate pressure turbine.

IP turbine : intermediate pressure turbine.

LPC : low pressure compressor.

LP compressor : low pressure compressor.

LPT : low pressure turbine.

LP turbine : low pressure turbine.

Net thrust : nozzle total gross thrust - intake ram drag (excluding nacelle drag, etc., this is the basic thrust acting on the airframe).

Overall pressure ratio : combustor inlet total pressure/intake delivery total pressure.

Overall thermal efficiency : thermal efficiency * propulsive efficiency.

Propulsive efficiency : propulsive power/rate of production of propulsive kinetic energy (maximum propulsive efficiency occurs when jet velocity equals flight velocity, which implies zero net thrust!).

SFC : Specific fuel consumption.

Specific fuel consumption : total fuel flow/net thrust (proportional to flight velocity/overall thermal efficiency).

Static pressure : normal meaning of pressure. Excludes any kinetic energy effects.

Specific thrust : net thrust/intake airflow.

Thermal efficiency : rate of production of propulsive kinetic energy/ fuel power.

Total fuel flow : combustor (plus any afterburner) fuel flow rate (e.g. lb/s or g/s).

Total pressure : static pressure plus kinetic energy term.

Turbine rotor inlet temperature : gas absolute mean temperature at principal (e.g. HP) turbine rotor entry.

Turboprop

A turboprop engine is a type of turbine engine which drives an aircraft propeller using a reduction gear.

The gas turbine is designed specifically for this application, with almost all of its output being used to drive the propeller. The engine's exhaust gases contain little energy compared to a jet engine and play only a minor role in the propulsion of the aircraft.

The propeller is coupled to the turbine through a reduction gear that converts the high RPM, low torque output to low RPM, high torque. The propeller itself is normally a constant speed (variable pitch) type similar to that used with larger reciprocating aircraft engines.

Turboprop engines are generally used on small subsonic aircraft, but some aircraft outfitted with turboprops have cruising speeds in excess of 500 kt (926 km/h, 575 mph). Large military and civil aircraft, such as the Lockheed L-188 Electra and the Tupolev Tu-95, have also used turboprop power. The Airbus A400M is powered by four Europrop TP400 engines, which are the third most powerful turboprop engines ever produced, after the Kuznetsov NK-12 and Progress D-27.

In its simplest form a turboprop consists of an intake, compressor, combustor, turbine, and a propelling nozzle. Air is drawn into the intake and compressed by the compressor. Fuel is then added to the compressed air in the combustor, where the fuel-air mixture then combusts. The hot combustion gases expand through the turbine. Some of the power generated by the turbine is used to drive the compressor. The rest is transmitted through the reduction gearing to the propeller. Further expansion of the gases occurs in the propelling nozzle, where the gases exhaust to atmospheric pressure. The propelling nozzle provides a relatively small proportion of the thrust generated by a turboprop.

Turboprops are very efficient at flight speeds below 450 mph because the jet velocity of the propeller (and exhaust) is relatively low. Due to the high price of turboprop engines, they are mostly used where high-performance short-takeoff and landing (STOL) capability and efficiency at modest flight speeds are required. The most common application of turboprop engines in civilian aviation is in small commuter aircraft, where their greater reliability than reciprocating engines offsets their higher initial cost. Turboprop airliners now operate at near the same speed as small turbofan-powered aircraft but burn two-thirds of the fuel per passenger. However, compared to a turbojet (which can fly at high altitude for enhanced speed and fuel consumption) a propeller aircraft has a much lower ceiling. Turboprop-powered aircraft have become popular for bush aeroplanes such as the Cessna Caravan and Quest Kodiak as jet fuel is easier to obtain in remote areas than is aviation-grade gasoline (avgas).

Technological Aspects

Much of the jet thrust in a turboprop is sacrificed in favour of shaft power, which is obtained by extracting additional power (up to that necessary to drive the compressor) from turbine expansion. While the power turbine may be integral with the gas generator section, many turboprops today feature a free power turbine on a separate coaxial shaft. This enables the propeller to rotate freely, independent of compressor speed. Owing to the additional expansion in the turbine system, the residual energy in the exhaust jet is low. Consequently, the exhaust jet produces (typically) less than 10% of the total thrust.

Propellers are not efficient when the tips reach or exceed supersonic speeds. For this reason, a reduction gearbox is placed in the drive line between the power turbine and the propeller to allow the turbine to operate at its most efficient speed while the propeller operates at its most efficient speed. The gearbox is part of the engine and contains the parts necessary to operate a constant speed propeller. This differs from the turboshaft engines used in helicopters, where the gearbox is remote from the engine.

Residual thrust on a turboshaft is avoided by further expansion in the turbine system and/or truncating and turning the exhaust 180 degrees, to produce two opposing jets. Apart from the above, there is very little difference between a turboprop and a turboshaft.

While most modern turbojet and turbofan engines use axial-flow compressors, turboprop engines usually contain at least one stage of centrifugal compression. Centrifugal compressors have the advantage of being simple and lightweight, at the expense of a streamlined shape.

Propellers lose efficiency as aircraft speed increases, so turboprops are normally not used on high-speed aircraft. However, propfan engines, which are very similar to turboprop engines, can cruise at flight speeds approaching Mach 0.75. To increase the efficiency of the propellers, a mechanism can be used to alter the pitch, thus adjusting the pitch to the airspeed. A variable pitch propeller, also called a controllable pitch propeller, can also be used to generate negative thrust while decelerating on the runway. Additionally, in the event of an engine outage, the pitch can be adjusted to a vaning pitch (called feathering), thus minimising the drag of the non-functioning propeller.

Some commercial aircraft with turboprop engines include the Bombardier Dash 8, ATR 42, ATR 72, BAe Jetstream 31, Embraer EMB 120 Brasilia, Fairchild Swearingen Metroliner, Saab 340 and 2000, Xian MA60, Xian MA600, and Xian MA700.

History

Alan Arnold Griffith had published a paper on turbine design in 1926. Subsequent work at the Royal Aircraft Establishment investigated axial turbine designs that could be used to supply power to a shaft and thence a propeller. From 1929, Frank Whittle began work on centrifugal turbine designs that would deliver pure jet thrust.

Gyorgy Jendrassik published a turboprop idea in 1928, on March 12, 1929 he patented his turboprop invention. In 1938, he built a small-scale (100 Hp) prototype of his patent. The world's first turboprop was the Jendrassik Cs-1, designed by the Hungarian mechanical engineer Gyorgy Jendrassik. It was produced and tested in the Ganz Works in Budapest between 1939 and 1942. It was planned to fit to the Varga RMI-1 X/H twin-engined reconnaissance bomber in 1940, but the program was cancelled.

The first British turboprop engine was the Rolls-Royce RB.50 Trent, a converted Derwent II fitted with reduction gear and a Rotol 7-ft, 11-in five-bladed propeller. Two Trents were fitted to Gloster Meteor *EE227* — the sole "Trent-Meteor" — which thus became the world's first turboprop-powered aircraft, albeit a test-bed not intended for production. It first flew on 20 September 1945. From their experience with the Trent, Rolls-Royce developed the Dart, which became one of the most reliable turboprop engines ever built. Dart production continued for more than fifty years. The Dart-powered Vickers Viscount was the first turboprop aircraft of any kind to go into production and sold in large numbers. It was also the first four-engined turboprop. Its first flight was on 16 July 1948. The world's first single engined turboprop aircraft was the Armstrong Siddeley Mamba-powered Boulton Paul Balliol, which first flew on 24 March 1948.

The Soviet Union built on German World War II development by Junkers (BMW and Hirth/Daimler-Benz also developed and partially tested designs). While the Soviet Union had the technology to create a jet-powered strategic bomber comparable to Boeing's B-52 Stratofortress, they instead produced the Tupolev Tu-95 Bear, powered with four Kuznetsov NK-12 turboprops, mated to eight contra-rotating propellers (two per nacelle) with supersonic tip speeds to achieve maximum cruise speeds in excess of 575 mph, faster than many of the first jet aircraft and comparable to jet cruising speeds for most missions. The Bear would serve as their most successful long-range combat and surveillance aircraft and symbol of Soviet power projection throughout the end of the 20th century.

The USA would incorporate contra-rotating turboprop engines, such as the ill-fated Allison T40, into a series of experimental aircraft during the 1950s, but none would be adopted into service.

The first American turboprop engine was the General Electric XT31, first used in the experimental Consolidated Vultee XP-81. The XP-81 first flew in December 1945, the first aircraft to use a combination of turboprop and turbojet power. The technology of the Lockheed Electra airliner was also used in military aircraft, such as the P-3 Orion and the C-130 Hercules, using the Allison T56. One of the most produced turboprop engines is the Pratt & Whitney Canada PT6 engine.

The first turbine-powered, shaft-driven helicopter was the Bell XH-13F, a version of the Bell 47 powered by Continental XT-51-T-3 (Turbomeca Artouste) engine.

Chapter 7

Turboshaft

A turboshaft engine is a form of gas turbine which is optimised to produce *free turbine* shaft power, rather than jet thrust. In concept, turboshaft engines are very similar to turbojets, with additional turbine expansion to extract heat energy from the exhaust and convert it into output shaft power.

Turboshaft engines are commonly used in applications which require a sustained high power output, high reliability, small size, and light weight. These include helicopters, auxiliary power units, boats and ships, tanks, hovercraft, and stationary equipment.

Overview

A turboshaft engine is made up of two major parts assemblies: the *gas generator* and the *power section*. The gas generator consists of the compressor, combustion chambers with ignitors and fuel nozzles, and one or more stages of turbine. The power section consists of additional stages of turbines, a gear reduction system, and the shaft output. The gas generator creates the hot expanding gases to drive the power section. Depending on the design, the engine accessories may be driven either by the gas generator or by the power section.

In most designs the gas generator and power section are mechanically separate so that they may each rotate at different speeds appropriate for the conditions. This is referred to as a *free power turbine*. A free power turbine can be an extremely useful design feature for vehicles, as it allows the design to forego the weight and cost of complex multi-ratio transmissions and clutches.

The general layout of a turboshaft is similar to that of a turboprop. The main difference is that a turboprop is structurally designed to support the loads created by a rotating propeller, as the propeller is not attached to anything but the engine itself. In contrast, turboshaft engines usually drive a transmission which is not structurally attached to the engine. The transmission is attached to the vehicle structure and supports the loads created instead of the engine. However, in practice many of the same engines are built in both turboprop and turboshaft versions, with only minor differences.

An unusual example of the turboshaft principle is the Pratt & Whitney F135-PW-600 engine for the STOVL F-35B - in conventional mode it operates as a turbofan, but when powering the LiftFan it switches partially to turboshaft mode to send power forward through a shaft (like a turboprop) and partially to turbojet mode to continue to send thrust to the rear nozzle.

History

The first gas turbine engines used for armoured fighting vehicle GT 101 was installed in Panther tank mid 1944. The first true turboshaft engine for helicopter was built by the French engine firm Turbomeca, led by the founder, Joseph Szydlowski. In 1948 they built the first French-designed turbine engine, the 100shp 782. Originally conceived as an auxiliary power unit (APU), it was soon adapted to aircraft propulsion, and found a niche as a powerplant for turboshaft-driven helicopters in the 1950s. In 1950 this work was used to develop the larger 280shp Artouste, which was widely used on the Aerospatiale Alouette II and other helicopters.

Propfan

A propfan was first defined as a small diametre, highly loaded multiple bladed variable pitch propulsor having swept blades with thin advanced airfoil sections, integrated with a nacelle contoured to retard the airflow through the blades thereby reducing compressibility losses and designed to operate with a turbine engine and using a single stage reduction gear resulting in a high performance. The propfan concept was first revealed by Carl Rohrbach and Bruce Metzger of the Hamilton Standard Division of United Technologies in 1975 and was patented by Robert Cornell and Carl Rohrbach of Hamilton Standard in 1979.

Later work by General Electric on similar propulsors was done under the name unducted fan, which was a modified turbofan engine, with the fan placed outside the engine nacelle on the same axis as the compressor blades. Propfans are also known as ultra-high bypass (UHB) engines and, most recently, open flux rotor jet engines. The design is intended to offer the speed and performance of a turbofan, with the fuel economy of a turboprop.

Limitations and Solutions

Propeller Blade Tip Speed Limit

Turboprops have an optimum speed below about 450 mph (700 km/h). The reason is that all propellers lose efficiency at high speed, due to an effect known as wave drag that occurs just below supersonic speeds. This powerful form of drag has a sudden onset, and led to the concept of a sound barrier when it was first encountered in the 1940s. In the case of a propeller, this effect can happen any time the propeller is spun fast enough that the blade tips near the speed of sound, even if the aircraft is motionless on the ground.

The most effective way to counteract this problem (to some degree) is by adding more blades to the propeller, allowing it to deliver more power at a lower rotational speed. This is why many World War II fighter designs started with two or three-blade propellers and by the end of the war were using up to five blades in some cases as the engines were upgraded and new propellers were needed to more effectively absorb that power.

The major downside to this approach is that adding blades makes the propeller harder to balance and maintain and the additional blades cause minor performance penalties (due to drag and efficiency issues). But even with these sorts of measures at some point the forward speed of the plane combined with the rotational speed of the propeller will once again result in wave drag problems. For most aircraft this will occur at speeds over about 450 mph (700 km/h).

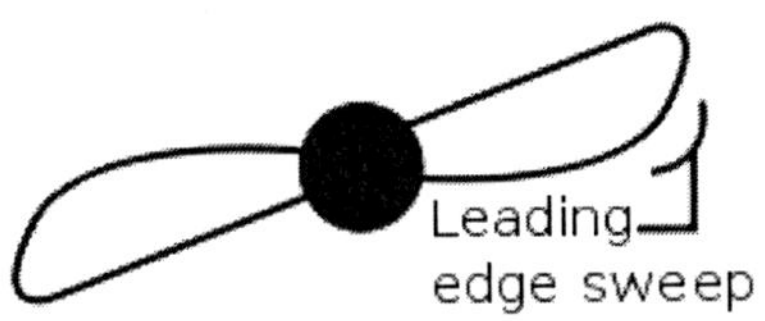

Figure 1: Swept propeller

A method of decreasing wave drag was discovered by German researchers in 1935—sweeping the wing backwards. Today, almost all aircraft designed to fly much above 450 mph (700 km/h) use a swept wing. In the 1970s, Hamilton Standard started researching propellers with similar sweep. Since the inside of the propeller is moving more slowly than the outside, the blade is progressively more swept toward the outside, leading to a curved shape similar to a scimitar - a practice that was first used as far back as 1909, in the Chauviere make of two-bladed wood propeller used on the Bleriot XI.

Jet Aircraft Fuel Economy

Jet aircraft are well known for permitting greater thrusts and higher speeds than could be achieved by conventional propeller-driven aircraft operating within the same aerodynamic envelope. However, jet aircraft are limited in fuel economy. In fact, for the same fuel consumption, a propeller-driven aircraft can produce greater thrust. As fuel costs become an increasingly important aspect of commercial aviation, aircraft engine designers continue to seek an optimal combination of jet engine thrust ratios and propeller fuel efficiency.

The propfan concept was developed to deliver 35% better fuel efficiency than contemporary turbofans. In static and air tests on a modified Douglas DC-9, propfans reached a 30% improvement over the OEM turbofans. This efficiency came at a price, as one of the major problems with the propfan is noise, particularly in an era where aircraft are required to comply with increasingly strict Stage III and Stage IV noise requirements.

The Hamilton Standard Division of United Technologies developed the propfan concept in the early 1970s. Numerous design variations of the propfan were tested by Hamilton Standard, in conjunction with NASA in this decade. This testing led to the Propfan Test Assessment (PTA) program, where Lockheed-Georgia proposed modifying a Gulfstream II to act as in-flight testbed for the propfan concept and McDonnell Douglas proposed modifying a DC-9 for the same purpose.

NASA choose the Lockheed proposal, where the aircraft had a nacelle added to the left wing, containing a 6000 hp Allison 570 turboprop engine, powering a 9-foot diametre Hamilton Standard SR-7 propfan. The aircraft, so configured, first flew in March 1987. After an extensive test program, the modifications were removed from the aircraft.

General Electric's GE36 Unducted Fan was a variation on the original propfan concept, and appears similar to a pusher configuration piston engine. GE's UDF has a novel direct drive arrangement, where the reduction gearbox is replaced by a low-speed seven-stage free turbine. The turbine rotors drive one propeller, while the other propeller is connected to the free turbine stators and rotates in the opposite direction. So, in effect, the power turbine has 14 stages. Boeing intended to offer GE's pusher UDF engine on the 7J7 platform, and McDonnell Douglas was going to do likewise on their MD-94X airliner. The GE36 was first flight tested mounted on the #3 engine station of a Boeing 727-100 in 1986.

McDonnell Douglas developed a proof-of-concept aircraft by modifying its company-owned MD-80. They removed the JT8D turbofan engine from the left side of the fuselage and replaced it with the GE36. A number of test flights were conducted, initially out of Mojave, California, which proved the airworthiness, aerodynamic characteristics, and noise signature of the design. Following the initial tests, a first-class cabin was installed inside the aft fuselage and airline executives were offered the opportunity to experience the UDF-powered aircraft first-hand. The test and marketing flights of the GE-outfitted demonstrator aircraft concluded in 1988, exhibiting a 30% reduction in fuel consumption over turbo-fan powered MD-80, full Stage III noise compliance, and low levels of interior noise/vibration. Due to jet fuel price drops and shifting marketing priorities, Douglas shelved the program the following year.

In the 1980s, Allison collaborated with Pratt & Whitney on demonstrating the 578-DX propfan. Unlike the competing GE36 UDF, the 578-DX was fairly conventional, having a reduction gearbox between the LP turbine and the propfan blades. The 578-DX was successfully flight tested on a McDonnell Douglas MD-80. However, none of the above projects came to fruition, mainly because of excessive cabin noise (compared to turbofans) and low fuel prices.

Figure 2: Progress D27 Propfans fitted to an Antonov An-70

The Progress D-27 propfan, developed in the U.S.S.R, is even more unconventional in layout, with the propfan blades at the front of the engine in a tractor configuration. Two rear-mounted D-27's propfans propelled the Antonov An-180, which was scheduled for a 1995 entry into service. Another Russian propfan application was the Yakovlev Yak-46. During the 1990s, Antonov also developed the An-70, powered by four Progress D-27s in a tractor configuration; the Russian Air Force placed an order for 164 aircraft in 2003, which was subsequently canceled. However, the An-70 remains available for further investment and production. With the currenthigh price for jet fuel and the emphasis on engine/airframe efficiency to reduce emissions, there is renewed interest in the propfan concept for jetliners that might come into service beyond the Boeing 787 and Airbus A350XWB. For instance, Airbus has patented aircraft designs with twin rear-mounted counter-rotating propfans.

Aircraft with Propfans

- Antonov An-70
- Beriev A-40
- EcoJet
- McDonnell Douglas MD-94X

Ramjet

A ramjet, sometimes referred to as a stovepipe jet, or an athodyd, is a form of airbreathing jet engine using the engine's forward motion to compress incoming air, without a rotary compressor. Ramjets cannot produce thrust at zero airspeed and thus cannot move an aircraft from a standstill.

Ramjets therefore require some other propulsion system to accelerate the vehicle to a speed where the ramjet begins to produce thrust. Ramjets require considerable forward speed to operate well, and as a class work most efficiently at speeds around Mach 3. This type of jet can operate up to speeds of Mach 6.

Ramjets can be particularly useful in applications requiring a small and simple engine for high speed use, such as missiles, while weapon designers are looking to use ramjet technology in artillery shells to give added range: it is anticipated that a 120-mm mortar shell, if assisted by a ramjet, could attain a range of 22 mi (35 km). They have also been used successfully, though not efficiently, as tip jets on helicopter rotors.

Ramjets are frequently confused with pulsejets, which use an intermittent combustion, but ramjets employ a continuous combustion process, and are a quite distinct type of jet engine.

History

Rene Lorin

The ramjet was invented in 1913 by French inventor Rene Lorin, who was granted a patent for his device. Attempts to build a prototype failed due to inadequate materials.

Albert Fono

In 1915 Hungarian inventor Albert Fono devised a solution for increasing the range of artillery, comprising a gun-launched projectile which was to be united with a ramjet propulsion unit, thus giving a long range from relatively low muzzle velocities, allowing heavy shells to be fired from relatively lightweight guns. Fono submitted his invention to the Austro-Hungarian Army but the proposal was rejected. After World War I, Fono returned to the subject of jet propulsion, in May 1928 describing an "air-jet engine" which he described as being suitable for high-altitude supersonic aircraft, in a German patent application. In an additional patent application he adapted the engine for subsonic speed. The patent was finally granted in 1932 after four years of examination.

Gorgon IV

The US Navy developed a series of air-to-air missiles under the name of "Gorgon" using different propulsion mechanisms, including ramjet propulsion. The ramjet Gorgon IVs, made by Glenn Martin, were tested in 1948 and 1949 at Naval Air Station Point Mugu. The ramjet engine itself was designed at the University of Southern California and manufactured by the Marquardt Aircraft Company. The engine was 7 feet long and 20 inches in diametre and was positioned below the missile.

Fritz Zwicky

The eminent Swiss astrophysicist was research director at Aerojet and holds many patents in jet propulsion. U.S. Patent 5121670 is for the Ram Accelerator and U.S. Patent 4722261 is the Extendable Ram Cannon. The U.S. Navy would not allow Fritz Zwicky to publicly discuss his own invention, U.S. Patent 2,461,797 for the Underwater Jet, a ram jet that performs in a fluid medium.TIME July 11, 1955 chronicles Fritz Zwicky's work in the "Missed Swiss" and the "Underwater Jet" in the March 14, 1949 issue.

Soviet Union

In the Soviet Union, a theory of supersonic ramjet engines was presented in 1928 by Boris S. Stechkin. Yuri Pobedonostsev, chief of GIRD's 3rd Brigade, carried out a great deal of research into ramjet engines. The first engine, the GIRD-04, was designed by I.A. Merkulov and tested in April 1933. To simulate supersonic flight, it was fed by air compressed to 200 atmospheres, and was fuelled with hydrogen. The GIRD-08 phosphorus-fuelled ramjet was tested by firing it from an artillery cannon. These shells may have been the first jet-powered projectiles to break the speed of sound.

In 1939, Merkulov did further ramjet tests using a two-stage rocket, the R-3.

In August of that year, he developed the first ramjet engine for use as an auxiliary motor of an aircraft, the DM-1. The world's first ramjet-powered aeroplane flight took place in December 1939, using two DM-2 engines on a modified Polikarpov I-15. Merkulov designed a ramjet fighter "Samolet D" in 1941, which was never completed. Two of his DM-4 engines were installed on the YaK-7 PVRD fighter, during World War II.

In 1940, the Kostikov-302 experimental plane was designed, powered by liquid fuel rocket for take-off and ramjet engines for flight. That project was cancelled in 1944.

In 1947, Mstislav Keldysh proposed a long-range antipodal bomber, similar to the Sänger-Bredt bomber, but powered by ramjet instead of rocket. In 1954, NPO Lavochkin and the Keldysh Institute began development of a trisonic ramjet-powered cruise missile, *Burya*. This project competed with the R-7 ICBM being developed by Sergei Korolev, and was cancelled in 1957.

Germany

In 1936 Hellmuth Walter constructed a test engine powered by natural gas. Theoretical work was carried out at BMW and Junkers as well as the DFL.

In 1941 Eugen Sänger of the DFL proposed a ramjet engine with a very high combustion chamber temperature. He constructed very large ramjet pipes with 500 millimetres (20 in) and 1,000 millimetres (39 in) diametre and carried out combustion tests on lorries and on a special test rig on a Dornier Do 17Z at flight speeds of up to 200 m/s (655 ft/s).

Later, with petrol becoming scarce in Germany due to wartime conditions, tests were carried out with blocks of pressed coal dust which were not successful due to slow combustion.

France

In France the works of Rene Leduc were notable. Leduc's Model, the Leduc 0.10 was one of the first ramjet-powered aircraft to fly, in 1949.

The Nord 1500 Griffon reached Mach 2.19 in 1958.

Engine Cycle

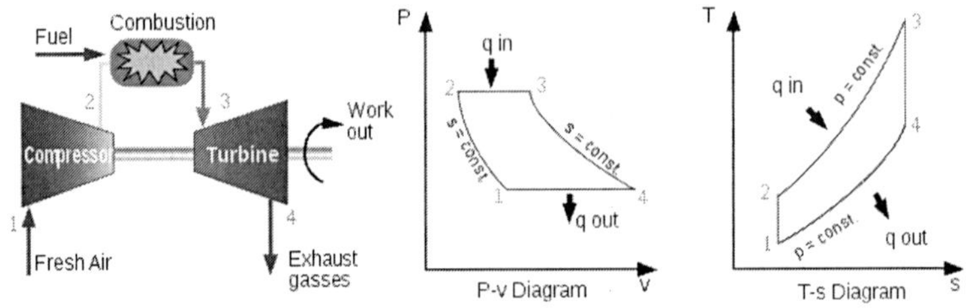

Figure 1: Brayton cycle

The Brayton cycle is a thermodynamic cycle that describes the workings of the gas turbine engine, basis of the airbreathing jet engine and others. It is named after George Brayton (1830–1892), the American engineer who developed it, although it was originally proposed and patented by Englishman John Barber in 1791. It is also sometimes known as the Joule cycle.

Design

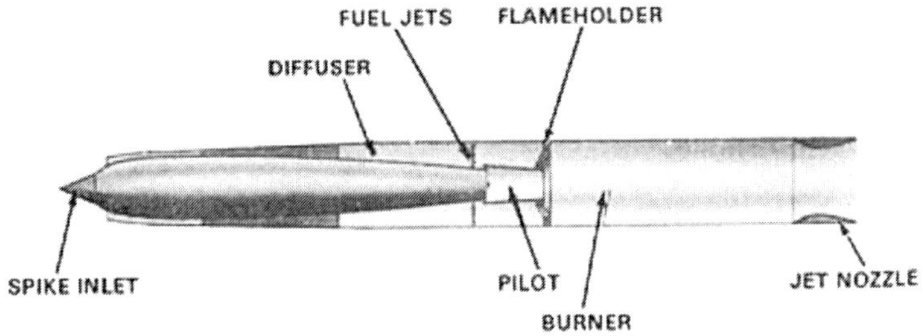

Figure 2: A typical ramjet

A ramjet is designed around its inlet. An object moving at high speed through air generates a high pressure region in front and a low pressure region to the rear. A ramjet uses this high pressure in front of the engine to force air through the tube, where it is heated by combusting some of it with fuel. It is then passed through a nozzle to accelerate it to supersonic speeds. This acceleration gives the ramjet forward thrust.

A ramjet is sometimes referred to as a 'flying stovepipe', a very simple device comprising an air intake, a combustor, and a nozzle. Normally the only moving parts are those within the turbopump, which pumps the fuel to the combustor in a liquid-fuel ramjet. Solid-fuel ramjets are even simpler.

By way of contrast, a turbojet uses a gas turbine driven fan to compress the air further. This gives greater compression and efficiency and far more power at low speeds, where the ram effect is weak, but is also more complex, heavier and expensive, and the temperature limits of the turbine section limit the top speed and thrust at high speed.

Inlet

Ramjets try to exploit the very high dynamic pressure within the air approaching the intake lip. An efficient intake will recover much of the freestream stagnation pressure, which is used to support the combustion and expansion process in the nozzle. Most ramjets operate at supersonic flight speeds and use one or more conical (or oblique) shock waves, terminated by a strong normal shock, to slow down the airflow to a subsonic velocity at the exit of the intake. Further diffusion is then required to get the air velocity down to a suitable level for the combustor.

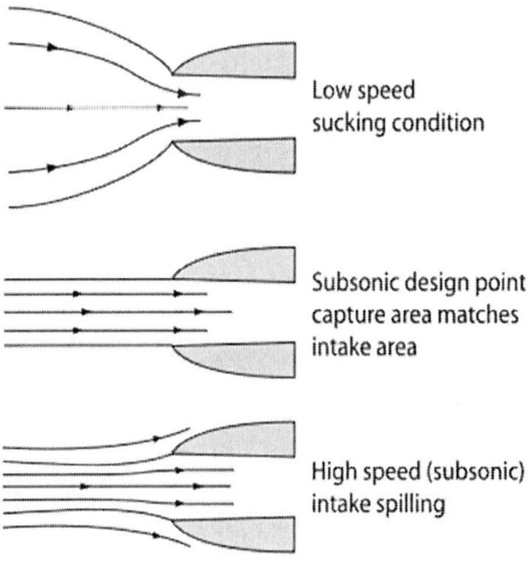

Figure 3: Subsonic intakes on ramjets are relatively simple

Subsonic ramjets do not need such a sophisticated inlet since the airflow is already subsonic and a simple hole is usually used. This would also work at slightly supersonic speeds, but as the air will choke at the inlet, this is inefficient. The Inlet is divergent, to provide a constant inlet speed of Mach 0.5.

Combustor

As with other jet engines the combustor's job is to create hot air. It does this by burning a fuel with the air at essentially constant pressure. The airflow through the jet engine is usually quite high, so sheltered combustion zones are produced by using *flame holders* that stop the flames from blowing out.

Since there is no downstream turbine, a ramjet combustor can safely operate at stoichiometric fuel: air ratios, which implies a combustor exit stagnation temperature of the order of 2400 K for kerosene. Normally the combustor must be capable of operating over a wide range of throttle settings, for a range of flight speeds/altitudes. Usually a sheltered pilot region enables combustion to continue when the vehicle intake undergoes high yaw/pitch, during turns. Other flame stabilisation techniques make use of flame holders, which vary in design from combustor cans to simple flat plates, to shelter the flame and improve fuel mixing. Overfuelling the combustor can cause the normal shock within a supersonic intake system to be pushed forward beyond the intake lip, resulting in a substantial drop in engine airflow and net thrust.

Nozzles

The propelling nozzle is a critical part of a ramjet design, since it accelerates exhaust flow to produce thrust.

For a ramjet operating at a subsonic flight Mach number, exhaust flow is accelerated through a converging nozzle. For a supersonic flight Mach number, acceleration is typically achieved via a convergent-divergent nozzle.

Figure 4: One of the two Bristol Thor ramjet engines on a Bristol Bloodhound missile

Performance and Control

Ramjets have been run from as low as 45 m/s (162 km/h) upwards. Below about Mach 0.5 they give little thrust and are highly inefficient due to their low pressure ratios. Above this speed, given sufficient initial flight velocity, a ramjet will be self-sustaining. Indeed, unless the vehicle drag is extremely high, the engine/airframe combination will tend to accelerate to higher and higher flight speeds, substantially increasing the air intake temperature.

As this could have a detrimental effect on the integrity of the engine and/or airframe, the fuel control system must reduce engine fuel flow to stabilise the flight Mach number and, thereby, air intake temperature to reasonable levels.

Due to the stoichiometric combustion temperature, efficiency is usually good at high speeds (Mach 2-3), whereas at low speeds the relatively poor compression ratio means that ramjets are outperformed by turbojets, or even rockets.

Ramjet Types

Ramjets can be classified according to the type of fuel, liquid or solid; and the booster.

In a liquid fuel ramjet (LFRJ) hydrocarbon fuel (typically) is injected into the combustor ahead of a flameholder which stabilises the flame resulting from the combustion of the fuel with the compressed air from the intake(s). A means of pressurising and supplying the fuel to the ramcombustor is required which can be complicated and expensive. Aerospatiale-Celerg have designed an LFRJ where the fuel is forced into the injectors by an elastomer bladder which inflates progressively along the length of the fuel tank. Initially the bladder forms a close-fitting sheath around the compressed air bottle from which it is inflated, which is mounted lengthwise in the tank. This offers a lower cost approach than a regulated LFRJ requiring a turbopump and associated hardware to supply the fuel.

A ramjet generates no static thrust and needs a booster to achieve a forward velocity high enough for efficient operation of the intake system. The first ramjet-powered missiles used external boosters, usually solid-propellant rockets, either in tandem, where the booster is mounted immediately aft of the ramjet, e.g. Sea Dart, or wraparound where multiple boosters are attached alongside the outside of the ramjet e.g. SA-4 Ganef. The choice of booster arrangement is usually driven by the size of the launch platform. A tandem booster increases the overall length of the system whereas wraparound boosters increase the overall diametre. Wraparound boosters will usually generate higher drag than a tandem arrangement.

Integrated boosters provide a more efficient packaging option since the booster propellant is cast inside the otherwise empty combustor. This approach has been used on solid, for example SA-6 Gainful, liquid, for example ASMP, and ducted rocket, for example Meteor, designs. Integrated designs are complicated by the different nozzle requirements of the boost and ramjet phases of flight. Due to the higher thrust levels of the booster a different shaped nozzle is required for optimum thrust compared to that required for the lower thrust ramjet sustainer. This is usually achieved via a separate nozzle which is ejected after booster burnout. However, designs such as Meteor feature nozzleless boosters. This offers the advantages of elimination of the hazard to launch aircraft from the ejected boost nozzle debris, simplicity, reliability, and reduced mass and cost, although this must be traded against the reduction in performance compared with that provided by a dedicated booster nozzle.

Integral Rocket Ramjet/Ducted Rocket

These are a slight variation on the ramjet where the supersonic exhaust from a rocket combustion process is used to compress and react with the incoming air in the main combustion chamber. This has the advantage of giving thrust even at zero speed. In a solid fuel integrated rocket ramjet (SFIRR) the solid fuel is cast along the outer wall of the ramcombustor. In this case fuel injection is through ablation of the propellant by the hot compressed air from the intake(s). An aft mixer may be used to improve combustion efficiency. SFIRRs are preferred over LFRJs for some applications because of the simplicity of the fuel supply but only when the throttling requirements are minimal i.e. when variations in altitude or Mach number are limited.

In a ducted rocket a solid fuel gas generator produces a hot fuel-rich gas which is burnt in the ramcombustor with the compressed air supplied by the intake(s). The flow of gas improves the mixing of the fuel and air and increases total pressure recovery. In a Throttleable Ducted Rocket (TDR), also known as a Variable Flow Ducted Rocket (VFDR), a valve allows the gas generator exhaust to be throttled allowing control of the thrust. Unlike an LFRJ solid propellant ramjets cannot flameout. The ducted rocket sits somewhere between the simplicity of the SFRJ and the unlimited throttleability of the LFRJ.

Flight Speed

Ramjets generally give little or no thrust below about half the speed of sound, and they are inefficient (less than 600 seconds) until the airspeed exceeds 1000 km/h (600 mph) due to low compression ratios. Even above the minimum speed a wide flight envelope (range of flight conditions), such as low to high speeds and low to high altitudes, can force significant design compromises, and they tend to work best optimised for one designed speed and altitude (point designs). However, ramjets generally outperform gas turbine based jet engine designs and work best at supersonic speeds (Mach 2–4). Although inefficient at slower speeds they are more fuel-efficient than rockets over their entire useful working range up to at least Mach 5.5.

The performance of conventional ramjets falls off above Mach 6 due to dissociation and pressure loss caused by shock as the incoming air is slowed to subsonic velocities for combustion. In addition, the combustion chamber's inlet temperature increases to very high values, approaching the dissociation limit at some limiting Mach number.

Related Engines—Air Turboramjet

Another example of this is the Air TurboRamjet (ATR) which has a compressor powered by a gas heated via a heat exchanger within the combustion chamber.

Scramjets

Ramjets always slow the incoming air to a subsonic velocity within the combustor. Scramjets, or "supersonic combustion ramjet" are similar to ramjets, but some of the air goes through the entire engine at supersonic speeds. This increases the stagnation pressure recovered from the freestream and improves net thrust. Thermal choking of the exhaust is avoided by having a relatively high supersonic air velocity at combustor entry. Fuel injection is often into a sheltered region below a step in the combustor wall. Although scramjet engines have been studied for many decades it is only recently that small experimental units have been flight tested and then only very briefly (e.g. the Boeing X-43).

As of May, 2010, this engine has been tested to attain Mach 5 for 200 seconds on the X-51A Waverider.

In the scramjet, the ram air is not slowed to subsonic speeds for combustion and as a result, shocks are not encountered and pressure loss is avoided.

Precooled Engines

A variant of the pure ramjet is the 'combined cycle' engine, intended to overcome the limitations of the pure ramjet. One example of this is the SABRE engine; this uses a precooler, behind which is ramjet and turbine machinery.

The ATREX engine developed in Japan is an experimental implementation of this concept. It uses liquid hydrogen fuel in a fairly exotic single-fan arrangement. The liquid hydrogen fuel is pumped through a heat exchanger in the air-intake, simultaneously heating the liquid hydrogen, and cooling the incoming air. This cooling of the incoming air is critical to achieving a reasonable efficiency. The hydrogen then continues through a second heat exchanger position after the combustion section, where the hot exhaust is used to further heat the hydrogen, turning it into a very high pressure gas. This gas is then passed through the tips of the fan providing driving power to the fan at sub-sonic speeds. After mixing with the air it is burned in the combustion chamber.

The Reaction Engines Scimitar has been proposed for the LAPCAT hypersonic airliner, and the Reaction Engines SABRE for the Reaction Engines Skylon spaceplane.

Nuclear Powered Ramjets

During the Cold War the United States designed and ground-tested a nuclear-powered ramjet called Project Pluto. This system used no combustion – a nuclear reactor heated the air instead. The project was ultimately canceled because ICBMs seemed to serve the purpose better, and because a low-flying radioactive missile could cause problems for any allied soldiers.

J58

The SR-71's Pratt & Whitney J58 engines act as turbojet-assisted ramjets at high-speeds (Mach 3.2).

Ionospheric Ramjet

The upper atmosphere above about 100 km contains monatomic oxygen that has been produced by the sun through photochemistry. A concept was created by NASA for recombining this thin gas back to diatomic molecules at orbital speeds to power a ramjet.

Bussard Ramjet

The Bussard ramjet is a space drive concept that is intended to fuse interstellar wind and exhaust it at high speed from the rear of the vehicle.

Scramjet

A scramjet (*supersonic combustion ramjet*) is a variant of a ramjet airbreathing jet engine in which combustion takes place in supersonic airflow. As in ramjets, a scramjet relies on high vehicle speed to forcefully compress and decelerate the incoming air before combustion (hence *ram*jet), but whereas a ramjet decelerates the air to subsonic velocities before combustion, airflow in a scramjet is supersonic throughout the entire engine. This allows the scramjet to efficiently operate at extremely high speeds: theoretical projections place the top speed of a scramjet between Mach 12 and Mach 24, which is near orbital velocity. The fastest air-breathing aircraft is a SCRAM jet design, the NASA X-43A which reached Mach 9.8. For comparison, the second fastest air-breathing aircraft, the manned SR-71 Blackbird, has a cruising speed of Mach 3.2.

The scramjet is composed of three basic components: a converging inlet, where incoming air is compressed and decelerated; a combustor, where gaseous fuel is burned with atmospheric oxygen to produce heat; and a diverging nozzle, where the heated air is accelerated to produce thrust. Unlike a typical jet engine, such as a turbojet or turbofan engine, a scramjet does not use rotating, fan-like components to compress the air; rather, the achievable speed of the aircraft moving through the atmosphere causes the air to compress within the inlet. As such, no moving parts are needed in a scramjet. In comparison, typical turbojet engines require inlet fans, multiple stages of rotating compressor fans, and multiple rotating turbine stages, all of which add weight, complexity, and a greater number of failure points to the engine.

Due to the nature of their design, scramjet operation is limited to near-hypersonic velocities. As they lack mechanical compressors, scramjets require the high kinetic energy of a hypersonic flow to compress the incoming air to operational conditions. Thus, a scramjet-powered vehicle must be accelerated to the required velocity by some other means of propulsion, such as turbojet, railgun, or rocket engines. In the flight of the experimental scramjet-powered Boeing X-51A, the test craft was lifted to flight altitude by a Boeing B-52 Stratofortress before being released and accelerated by a detachable rocket to near Mach 4.5.

While scramjets are conceptually simple, actual implementation is limited by extreme technical challenges. Hypersonic flight within the atmosphere generates immense drag, and temperatures found on the aircraft and within the engine can be much greater than that of the surrounding air. Maintaining combustion in the supersonic flow presents additional challenges, as the fuel must be injected, mixed, ignited, and burned within milliseconds. While scramjet technology has been under development since the 1950s, only very recently have scramjets successfully achieved powered flight.

History

During and after World War II, tremendous amounts of time and effort were put into researching high-speed jet- and rocket-powered aircraft. The Bell X-1 attained supersonic flight in 1947 and, by the early 1960s, rapid progress towards faster aircraft suggested that operational aircraft would be flying at "hypersonic" speeds within a few years. Except for specialised rocket research vehicles like the North American X-15 and other rocket-powered spacecraft, aircraft top speeds have remained level, generally in the range of Mach 1 to Mach 3.

In the 1950s and 1960s a variety of experimental scramjets engines were built and ground tested in US and the UK. In 1964, Dr. Frederick S. Billig and Dr. Gordon L. Dugger submitted a patent application for a supersonic combustion ramjet based on Billig's Ph.D. thesis. This patent was issued in 1981 following the removal of an order of secrecy.

In 1981 tests were made in Australia under the guidance of Professor Ray Stalker in the T3 ground test facility at ANU.

First successful flight test of Scramjet was performed by Russia in 1991. It was axisymmetric hydrogen-fuelled dual-mode scramjet developed by Central Institute of Aviation Motors (CIAM), Moscow from late 1970s. The scramjet fight was known captive-carry atop the SA-5 surface-to-air missile that included an experiment fight support unit known as the "Hypersonic Flying Laboratory" (HFL), "Kholod". Then during the period since 1992 up to 1998 an additional 6 flight tests of the axisymmetric high-speed scramjet-demonstrator were conducted by CIAM together with France and then with NASA, USA. Maximum flight velocity greater than Mach 6.4 was achieved and Scramjet operation during 77 seconds was demonstrated. These flight test series also provided insight into autonomous hypersonic flight controls.

Investigations of hypersonic flight have continued over the past few decades. At present, the US military and NASA have formulated a "National Hypersonics Strategy" to investigate a range of options for hypersonic flight.

Different U.S. organisations have accepted hypersonic flight as a common goal. The U.S. Army desires hypersonic missiles that can attack mobile missile launchers quickly. NASA believes hypersonics could help develop economical, reusable launch vehicles. The Air

Force is interested in a range of hypersonic systems, from air-launched cruise missiles to orbital spaceplanes, that the service believes could bring about a true "aerospace force."

The problem is complicated by the release of previously classified material and by partial publication, where claims are made, but specific parts of an experiment are kept secret. Additionally experimental difficulties in verifying that supersonic combustion actually occurred, or that actual net thrust was produced mean that at least four consortia have legitimate claims to "firsts", with several nations and institutions involved in each consortium. On June 15, 2007, the US Defence Advanced Research Project Agency (DARPA) and the Australian Defence Science and Technology Organisation (DSTO), announced a successful scramjet flight at Mach 10 using rocket engines to boost the test vehicle to hypersonic speeds, at the Woomera Rocket Range in Central Australia.

No scramjet powered vehicle has yet been produced outside an experimental program.

Design Principles

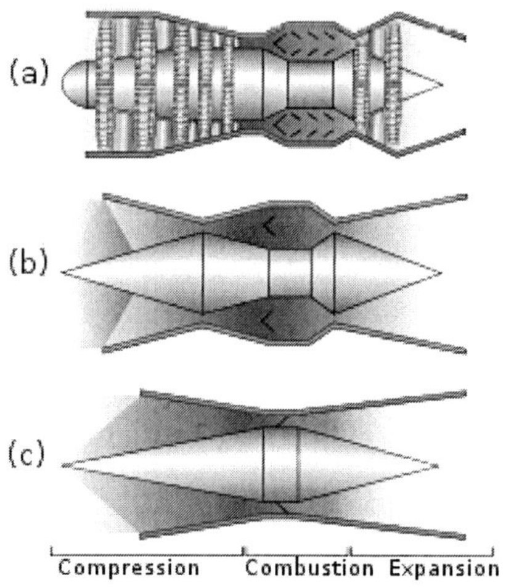

Figure 5: The compression, combustion, and expansion regions of: (a) turbojet, (b) ramjet, and (c) scramjet engines.

Scramjet engines are a type of jet engine, and rely on the combustion of fuel and an oxidiser to produce thrust. Similar to conventional jet engines, scramjet-powered aircraft carry the fuel on board, and obtain the oxidiser by the ingestion of atmospheric oxygen (as compared to rockets, which carry both fuel and an oxidising agent).

This requirement limits scramjets to suborbital atmospheric flight, where the oxygen content of the air is sufficient to maintain combustion.

Basic Principles

Scramjets are designed to operate in the hypersonic flight regime, beyond the reach of turbojet engines, and, along with ramjets, fill the gap between the high efficiency of turbojets and the high speed of rocket engines. Turbomachinery-based engines, while highly efficient at subsonic speeds, become increasingly inefficient at transonic speeds, as the compressor fans found in turbojet engines require subsonic speeds to operate. While the flow from transonic to low supersonic speeds can be decelerated to these conditions, doing so at supersonic speeds results in a tremendous increase in temperature and a loss in the total enthalpy of the flow. Around Mach 3–4, turbomachinery is no longer useful, and ram-style compression becomes the preferred method.

Ramjets utilise high-speed characteristics of air to literally 'ram' air through an inlet nozzle into the combustor. At transonic and supersonic flight speeds, the air upstream of the nozzle is not able to move out of the way quickly enough, and is compressed within the nozzle before being diffused into the combustor. Combustion in a ramjet takes place at subsonic velocities, similar to turbojets, but the combustion products are then accelerated through a convergent-divergent nozzle to supersonic speeds. As they have no mechanical means of compression, ramjets cannot start from a standstill, and generally do not achieve sufficient compression until supersonic flight. The lack of intricate turbomachinery allows ramjets to deal with the temperature rise associated with decelerating a supersonic flow to subsonic speeds, but this only goes so far: at near-hypersonic velocities, the temperature rise and inefficiencies discourage decelerating the flow to the magnitude found in ramjet engines.

Scramjet engines operate on the same principles as ramjets, but do not decelerate the flow to subsonic velocities. Rather, a scramjet combustor is supersonic: the inlet decelerates the flow to a lower Mach number for combustion, after which it is accelerated to an even higher Mach number through the nozzle. By limiting the amount of deceleration, temperatures within the engine are kept at a tolerable level, from both a material and combustive standpoint. Even so, current scramjet technology requires the use of high-energy fuels and active cooling schemes to maintain sustained operation, often using hydrogen and regenerative cooling techniques.

Theory

All scramjet engines have an intake, which compresses the incoming air, fuel injectors, a combustion chamber, and a divergent thrust nozzle. Sometimes engines also include a region which acts as a flame holder, although the high stagnation temperatures mean that an area of focused waves may be used, rather than a discrete engine part as seen in turbine engines. Other engines use pyrophoric fuel additives, such as silane, to avoid such issues. An isolator between the inlet and combustion chamber is often included to improve the homogeneity of the flow in the combustor and to extend the operating range of the engine.

Figure 6: Computational fluid dynamics (CFD) image of the NASA X-43A with scramjet attached to the underside at Mach 7

A scramjet is reminiscent of a ramjet. In a typical ramjet, the supersonic inflow of the engine is decelerated at the inlet to subsonic speeds and then reaccelerated through a nozzle to supersonic speeds to produce thrust. This deceleration, which is produced by a normal shock, creates a total pressure loss which limits the upper operating point of a ramjet engine.

For a scramjet, the kinetic energy of the freestream air entering the scramjet engine is large comparable to the energy released by the reaction of the oxygen content of the air with a fuel (say hydrogen). Thus the heat released from combustion at Mach 25 is around 10% of the total enthalpy of the working fluid. Depending on the fuel, the kinetic energy of the air and the potential combustion heat release will be equal at around Mach 8. Thus the design of a scramjet engine is as much about minimising drag as maximising thrust.

This high speed makes the control of the flow within the combustion chamber more difficult. Since the flow is supersonic, no upstream influence propagates within the freestream of the combustion chamber.

Thus throttling of the entrance to the thrust nozzle is not a usable control technique. In effect, a block of gas entering the combustion chamber must mix with fuel and have sufficient time for initiation and reaction, all the while travelling supersonically through the combustion chamber, before the burned gas is expanded through the thrust nozzle. This places stringent requirements on the pressure and temperature of the flow, and requires that the fuel injection and mixing be extremely efficient. Usable dynamic pressures lie in the range 20 to 200 kPa (0.2-2 bar), where

$$q = \frac{1}{2}\rho v^2$$

where

q is the dynamic pressure of the gas

ρ (rho) is the density of the gas

v is the velocity of the gas

To keep the combustion rate of the fuel constant, the pressure and temperature in the engine must also be constant. This is problematic because the airflow control systems that would facilitate this are not physically possible in a scramjet launch vehicle due to the large speed and altitude range involved, meaning that it must travel at an altitude specific to its speed. Because air density reduces at higher altitudes, a scramjet must climb at a specific rate as it accelerates to maintain a constant air pressure at the intake. This optimal climb/descent profile is called a "constant dynamic pressure path". It is thought that scramjets might be operable up to an altitude of 75 km.

Fuel injection and management is also potentially complex. One possibility would be that the fuel be pressurised to 100 bar by a turbo pump, heated by the fuselage, sent through the turbine and accelerated to higher speeds than the air by a nozzle. The air and fuel stream are crossed in a comb like structure, which generates a large interface. Turbulence due to the higher speed of the fuel leads to additional mixing. Complex fuels like kerosene need a long engine to complete combustion.

The minimum Mach number at which a scramjet can operate is limited by the fact that the compressed flow must be hot enough to burn the fuel, and have pressure high enough that the reaction be finished before the air moves out the back of the engine. Additionally, in order to be called a scramjet, the compressed flow must still be supersonic after combustion. Here two limits must be observed: Firstly, since when a supersonic flow is compressed it slows down, the level of compression must be low enough (or the initial speed high enough) not to slow the gas below Mach 1. If the gas within a scramjet goes below Mach 1 the engine will "choke", transitioning to subsonic flow in the combustion chamber. This effect is well known amongst experimenters on scramjets since the waves caused by choking are easily observable. Additionally, the sudden increase in pressure and temperature in the engine can lead to an acceleration of the combustion, leading to the combustion chamber exploding.

Secondly, the heating of the gas by combustion causes the speed of sound in the gas to increase (and the Mach number to decrease) even though the gas is still travelling at the same speed. Forcing the speed of air flow in the combustion chamber under Mach 1 in this way is called "thermal choking". It is clear that a pure scramjet can operate at Mach numbers of 6-8, but in the lower limit, it depends on the definition of a scramjet. Certainly there are designs where a ramjet transforms into a scramjet over the Mach 3-6 range (Dual-mode scramjets). In this range however, the engine is still receiving significant thrust from subsonic combustion of "ramjet" type.

The high cost of flight testing and the unavailability of ground facilities have hindered scramjet development. A large amount of the experimental work on scramjets has been undertaken in cryogenic facilities, direct-connect tests, or burners, each of which simulates one aspect of the engine operation. Further, vitiated facilities, storage heated facilities, arc facilities and the various types of shock tunnels each have limitations which have prevented perfect simulation of scramjet operation. The HyShot flight test showed the relevance of the 1:1 simulation of conditions in the T4 and HEG shock tunnels, despite having cold models and a short test time. The NASA-CIAM tests provided similar verification for CIAM's C-16 V/K facility and the Hyper-X project is expected to provide similar verification for the Langley AHSTF, CHSTF and 8 ft (2.4 m) HTT.

Computational fluid dynamics has only recently reached a position to make reasonable computations in solving scramjet operation problems. Boundary layer modelling, turbulent mixing, two-phase flow, flow separation, and real-gas aerothermodynamics continue to be problems on the cutting edge of CFD. Additionally, the modelling of kinetic-limited combustion with very fast-reacting species such as hydrogen makes severe demands on computing resources. Reaction schemes are numerically stiff requiring reduced reaction schemes.

Much of scramjet experimentation remains classified. Several groups including the US Navy with the SCRAM engine between 1968–1974, and the Hyper-X program with the X-43A have claimed successful demonstrations of scramjet technology. Since these results have not been published openly, they remain unverified and a final design method of scramjet engines still does not exist.

The final application of a scramjet engine is likely to be in conjunction with engines which can operate outside the scramjet's operating range. Dual-mode scramjets combine subsonic combustion with supersonic combustion for operation at lower speeds, and rocket-based combined cycle (RBCC) engines supplement a traditional rocket's propulsion with a scramjet, allowing for additional oxidiser to be added to the scramjet flow. RBCCs offer a possibility to extend a scramjet's operating range to higher speeds or lower intake dynamic pressures than would otherwise be possible.

Advantages and Disadvantages of Scramjets

Advantages

1. Does not have to carry oxygen
2. No rotating parts makes it easier to manufacture
3. Has a higher specific impulse (change in momentum per unit of propellant) than a conventional engine; could provide between 1000 and 4000 seconds, while a rocket only provides 600 seconds or less
4. Higher speed could mean cheaper access to outer space in the future.

Special Cooling and Materials

Unlike a rocket that quickly passes mostly vertically through the atmosphere or a turbojet or ramjet that flies at much lower speeds, a hypersonic airbreathing vehicle optimally flies a "depressed trajectory", staying within the atmosphere at hypersonic speeds. Because scramjets have only mediocre thrust-to-weight ratios, acceleration would be limited. Therefore time in the atmosphere at hypersonic speed would be considerable, possibly 15–30 minutes. Similar to a reentering space vehicle, heat insulation would be a formidable task. The time in the atmosphere would be greater than that for a typical space capsule, but less than that of the space shuttle.

New materials offer good insulation at high temperature, but they often sacrifice themselves in the process. Therefore studies often plan on "active cooling", where coolant circulating throughout the vehicle skin prevents it from disintegrating. Often the coolant is the fuel itself, in much the same way that modern rockets use their own fuel and oxidiser as coolant for their engines. All cooling systems add weight and complexity to a launch system. The cooling of scramjets in this way may result in greater efficiency, as heat is added to the fuel prior to entry into the engine, but result in increased complexity and weight which ultimately could outweigh any performance gains.

Vehicle Performance

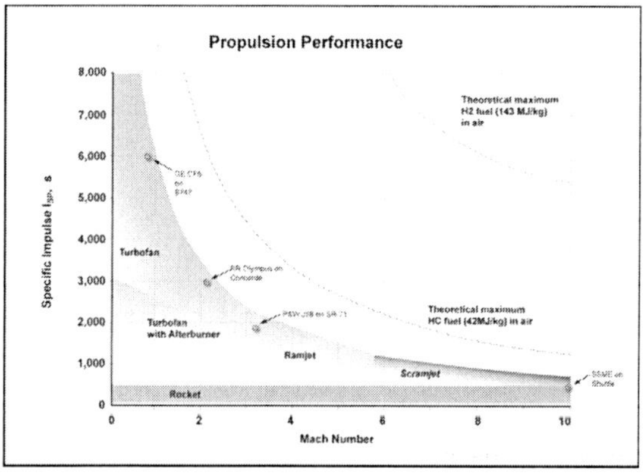

Figure 7: The specific impulse of various engines

The performance of a launch system is complex and depends greatly on its weight. Normally craft are designed to maximise range (R), orbital radius (R) or payload mass fraction (Γ) for a given engine and fuel. This results in tradeoffs between the efficiency of the engine (takeoff fuel weight) and the complexity of the engine (takeoff dry weight), which can be expressed by the following:

$$\Pi_e + \Pi_f + \frac{1}{\Gamma} = 1$$

Where :

- $\Pi_e = \frac{m_{\text{empty}}}{m_{\text{initial}}}$ is the empty mass fraction, and represents the weight of the superstructure, tankage and engine.
- $\Pi_f = \frac{m_{\text{fuel}}}{m_{\text{initial}}}$ is the fuel mass fraction, and represents the weight of fuel, oxidiser and any other materials which are consumed during the launch.

- $\Gamma = \dfrac{m_{\text{initial}}}{m_{\text{payload}}}$ is initial mass ratio, and is the inverse of the payload mass fraction. This represents how much payload the vehicle can deliver to a destination.

A scramjet increases the mass of the engine Π_e over a rocket, and decreases the mass of the fuel Π_f. It can be difficult to decide whether this will result in an increased Γ (which would be an increased payload delivered to a destination for a constant vehicle takeoff weight). The logic behind efforts driving a scramjet is (for example) that the reduction in fuel decreases the total mass by 30%, while the increased engine weight adds 10% to the vehicle total mass. Unfortunately the uncertainty in the calculation of any mass or efficiency changes in a vehicle is so great that slightly different assumptions for engine efficiency or mass can provide equally good arguments for or against scramjet powered vehicles.

Additionally, the drag of the new configuration must be considered. The drag of the total configuration can be considered as the sum of the vehicle drag (D) and the engine installation drag (D_e). The installation drag traditionally results from the pylons and the coupled flow due to the engine jet, and is a function of the throttle setting. Thus it is often written as:

$D_e = \phi_e F$ Where:

- ϕ_e is the loss coefficient
- F is the thrust of the engine

For an engine strongly integrated into the aerodynamic body, it may be more convenient to think of (D_e) as the difference in drag from a known base configuration.

The overall engine efficiency can be represented as a value between 0 and 1 (η_0), in terms of the specific impulse of the engine:

$$\eta_0 = \frac{g_0 V_0}{h_{PR}} \cdot I_{sp} = \frac{\text{Thrust Power}}{\text{Chemical energy rate}}$$

Where:

- g_0 is the acceleration due to gravity at ground level
- V_0 is the vehicle speed
- I_{sp} is the specific impulse
- h_{PR} is fuel heat of reaction

Specific impulse is often used as the unit of efficiency for rockets, since in the case of the rocket, there is a direct relation between specific impulse, specific fuel consumption and exhaust velocity. This direct relation is not generally present for airbreathing engines, and so specific impulse is less used in the literature. Note that for an airbreathing engine, both η_0 and I_{sp} are a function of velocity.

The specific impulse of a rocket engine is independent of velocity, and common values are between 200 and 600 seconds (450s for the space shuttle main engines). The specific impulse of a scramjet varies with velocity, reducing at higher speeds, starting at about 1200s, although values in the literature vary.

For the simple case of a single stage vehicle, the fuel mass fraction can be expressed as:

$$\Pi_f = 1 - \exp\left[-\frac{\left(\frac{V_{initial}^2}{2} - \frac{V_i^2}{2}\right) + \int g\,dr}{\eta_0 h_{PR}\left(1 - \frac{D + D_e}{F}\right)}\right]$$

Where this can be expressed for single stage transfer to orbit as:

$$\Pi_f = 1 - \exp\left[-\frac{g_0 r_0\left(1 - \frac{1}{2}\frac{r_0}{r}\right)}{\eta_0 h_{PR}\left(1 - \frac{D + D_e}{F}\right)}\right]$$

or for level atmospheric flight from air launch (missile flight):

$$\Pi_f = 1 - \exp\left[-\frac{g_0 R}{\eta_0 h_{PR}\left(1 - \phi_e\right)\frac{C_L}{C_D}}\right]$$

Where R is the range, and the calculation can be expressed in the form of the Breguet range formula:

$$\Pi_f = 1 - e^{-BR}$$

$$B = \frac{g_0}{\eta_0 h_{PR}\left(1 - \phi_e\right)\frac{C_L}{C_D}}$$

Where:

- C_L is the lift coefficient
- C_D is the drag coefficient

This extremely simple formulation, used for the purposes of discussion assumes:

- Single stage vehicle
- No aerodynamic lift for the transatmospheric lifter

However they are true generally for all engines.

Initial Propulsion Requirements

A scramjet cannot produce efficient thrust unless boosted to high speed, around Mach 5, although depending on the design it could act as a ramjet at low speeds. A horizontal take-off aircraft would need conventional turbofan or rocket engines to take off, sufficiently large to move a heavy craft. Also needed would be fuel for those engines, plus all engine-associated mounting structure and control systems. Turbofan engines are heavy and cannot easily exceed about Mach 2-3, so another propulsion method would be needed to reach scramjet operating speed.

That could be ramjets or rockets. Those would also need their own separate fuel supply, structure, and systems. Many proposals instead call for a first stage of droppable solid rocket boosters, which greatly simplifies the design.

Testing Difficulties

Figure 8: Test of Pratt & Whitney Rocketdyne SJY61 scramjet engine for the Boeing X-51

Unlike jet or rocket propulsion systems facilities which can be tested on the ground, testing scramjet designs use extremely expensive hypersonic test chambers or expensive launch vehicles, both of which lead to high instrumentation costs. Tests using launched test vehicles very typically end with destruction of the test item and instrumentation.

Advantages and Disadvantages for Orbital Vehicles

Propellant

An advantage of a hypersonic airbreathing (typically scramjet) vehicle like the X-30 is avoiding or at least reducing the need for carrying oxidiser. For example the space shuttle external tank holds 616,432 kg of liquid oxygen (LOX) and 103,000 kg of liquid hydrogen (LH2) while having an empty weight of 30,000 kg. The orbiter gross weight is 109,000 kg with a maximum payload of about 25,000 kg and to get the assembly off the launch pad the shuttle uses two very powerful solid rocket boosters with a weight of 590,000 kg each. If the oxygen could be eliminated, the vehicle could be lighter at liftoff and maybe carry more payload. That would be an advantage, but the central motivation in pursuing hypersonic airbreathing vehicles would be to reduce cost.

On the other hand, scramjets spend more time in the atmosphere and require more hydrogen fuel to deal with aerodynamic drag. Whereas liquid oxygen is quite a dense fluid, liquid hydrogen has much lower density and takes up much more volume. This means that the vehicle using this fuel becomes much bigger and gives even more drag.

Thrust-to-weight Ratio

One issue is that scramjet engines are predicted to have exceptionally poor thrust-to-weight ratio, when installed in a launch vehicle, of around 2. A rocket has the advantage that its engines have *very* high thrust-weight ratios (~100:1), while the tank to hold the liquid oxygen approaches a tankage ratio of ~100:1 also. Thus a rocket can achieve a very high mass fraction, which improves performance. By way of contrast the projected thrust/weight ratio of scramjet engines of about 2 mean a very much larger percentage of the takeoff mass is engine. In addition the vehicle's lower thrust does not necessarily avoid the need for the expensive, bulky, and failure prone high performance turbopumps found in conventional liquid-fuelled rocket engines, since most scramjet designs seem to be incapable of orbital speeds in airbreathing mode, and hence extra rocket engines are needed.

Need for Additional Propulsion to Reach Orbit

Scramjets might be able to accelerate from approximately Mach 5-7 to around somewhere between half of orbital velocity and orbital velocity. Generally, another propulsion system (very typically rocket is proposed) is expected to be needed for the final acceleration into orbit. Since the delta-V is moderate and the payload fraction of scramjets high, lower performance rockets such as solids, hypergolics, or simple liquid fuelled boosters might be acceptable.

Reentry

The scramjet's heat-resistant underside potentially doubles as its reentry system if a single-stage-to-orbit vehicle using non-ablative, non-active cooling is visualised. If an ablative shielding is used on the engine it will probably not be usable after ascent to orbit. If active cooling is used the loss of all fuel during the burn to orbit will also mean the loss of all cooling for the thermal protection system.

Costs

Reducing the amount of fuel and oxidiser does not necessarily improve costs as rocket propellants are comparatively very cheap. Indeed, the unit cost of the vehicle can be expected to end up far higher, since aerospace hardware cost is about two orders of magnitude higher than liquid oxygen, fuel and tankage, and scramjet hardware seems to be much heavier than rockets for any given payload. Still, if scramjets enable reusable vehicles, this could theoretically be a cost benefit. Whether equipment subject to the extreme conditions of a scramjet can be reused sufficiently many times is unclear; all flown scramjet tests only survive for short periods and have never been designed to survive a flight to date.

The eventual cost of such a vehicle is the subject of intense debate since even the best estimates disagree whether a scramjet vehicle would be advantageous. It is likely that a scramjet vehicle would need to lift more load than a rocket of equal takeoff weight in order to be equally as cost efficient (if the scramjet is a non-reusable vehicle).

Issues

Space launch vehicles may or may not benefit from having a scramjet stage. A scramjet stage of a launch vehicle theoretically provides a specific impulse of 1000 to 4000 s whereas a rocket provides less than 450 s while in the atmosphere, potentially permitting much cheaper access to space. A scramjet's specific impulse decreases rapidly with speed, however, and the vehicle would suffer from a relatively low lift to drag ratio.

The installed thrust to weight ratio of scramjets compares very unfavourably with the 50-100 of a typical rocket engine. This is compensated for in scramjets partly because the weight of the vehicle would be carried by aerodynamic lift rather than pure rocket power (giving reduced 'gravity losses'), but scramjets would take much longer to get to orbit due to lower thrust which greatly offsets the advantage. The takeoff weight of a scramjet vehicle is significantly reduced over that of a rocket, due to the lack of onboard oxidiser, but increased by the structural requirements of the larger and heavier engines.

Whether this vehicle would be reusable or not is still a subject of debate and research.

Applications

An aircraft using this type of jet engine could dramatically reduce the time it takes to travel from one place to another, potentially putting any place on Earth within a 90 minute flight. However, there are questions about whether such a vehicle could carry enough fuel to make useful length trips, and there are issues with sonic booms.

Scramjet vehicle has been proposed for a single stage to tether vehicle, where a Mach 12 spinning orbital tether would pick up a payload from a vehicle at around 100 km and carry it to orbit.

Progress in the 2000s

In the 2000s, significant progress was made in the development of hypersonic technology, particularly in the field of scramjet engines.

US efforts are probably the best funded, and the Hyper-X team claimed the first flight of a thrust-producing scramjet powered vehicle with full aerodynamic maneuvering surfaces in 2004 with the X-43A. However, the first group to demonstrate a scramjet working in an atmospheric test was a project by a joint British and Australian team from UK defence company QinetiQ and the University of Queensland.

The HyShot project demonstrated scramjet combustion on July 30, 2002. The scramjet engine worked effectively and demonstrated supersonic combustion in action. However, the engine was not designed to provide thrust to propel a craft. It was designed more or less as a technology demonstrator.

On Friday, June 15, 2007, the US Defence Advanced Research Project Agency (DARPA), in cooperation with the Australian Defence Science and Technology Organisation (DSTO), announced a successful scramjet flight at Mach 10 using rocket engines to boost the test vehicle to hypersonic speeds.

A series of scramjet ground test were completed at NASA Langley Arc-Heated Scramjet Test Facility (AHSTF) at simulated Mach 8 flight conditions. These experiments were used to support HiFIRE flight 2.

On May 22, 2009, Woomera hosted the first successful test flight of a hypersonic aircraft in HIFire. The launch was one of 10 planned test flights. The series of up to 10 planned hypersonic flight experiments are part of a joint research program between the Defence Science and Technology Organisation and the US Air Force, designated as the Hypersonic International Flight Research Experimentation (HIFiRE). HIFiRE is investigating hypersonics technology (the study of flight exceeding five times the speed of sound) and its application to advanced scramjet-powered space launch vehicles — the objective is to support the new Boeing X-51 scramjet demonstrator while also building a strong base of flight test data for quick-reaction space launch development and hypersonic "quick-strike" weapons.

On 22 and 23 March 2010, Australian and American defence scientists successfully tested a (HIFiRE) hypersonic rocket. It reached an atmospheric velocity of "more than 5,000 kilometres per hour" after taking off from the Woomera Test Range in outback South Australia.

On May 27, 2010, NASA and the United States Air Force successfully flew the X-51A Waverider for approximately 200 seconds at Mach 5, setting a new world record hypersonic airspeed. The Waverider flew autonomously before losing acceleration for an unknown reason and destroying itself as planned. The test was declared a success. The X-51A was carried aboard a B-52, accelerated to Mach 4.5 via a solid rocket booster, and then ignited the Pratt & Whitney Rocketdyne scramjet engine to reach Mach 5 at 70,000 feet. However, a second flight June 13, 2011 was ended prematurely when the engine lit briefly on ethylene but failed to transition to its primary JP7 fuel, failing to reach full power.

On 16 November 2010, Australian scientists successfully demonstrated that the high-speed flow in a naturally non-burning scramjet engine can be ignited using a pulsed laser source.

The US Army successfully tested the scramjet powered "Advanced Hypersonic Weapon" (AHW) on November 18, 2011. The weapon reached speeds over Mach 5 and hit a target around 2500 miles distant.

Combustion

Combustion or burning is the sequence of exothermic chemical reactions between a fuel and an oxidant accompanied by the production of heat and conversion of chemical species. The release of heat can result in the production of light in the form of either glowing or a flame. Fuels of interest often include organic compounds (especially hydrocarbons) in the gas, liquid or solid phase.

In a complete combustion reaction, a compound reacts with an oxidising element, such as oxygen or fluorine, and the products are compounds of each element in the fuel with the oxidising element. For example:

$$CH_4 + 2\ O_2 \rightarrow CO_2 + 2\ H_2O + \text{energy}$$

$$CH_2S + 6\ F_2 \rightarrow CF_4 + 2\ HF + SF_6$$

A simple example can be seen in the combustion of hydrogen and oxygen, which is a commonly used reaction in rocket engines:

$$2\ H_2 + O_2 \rightarrow 2\ H_2O(g) + \text{heat}$$

The result is water vapour.

Complete combustion is almost impossible to achieve. In reality, as actual combustion reactions come to equilibrium, a wide variety of major and minor species will be present such as carbon monoxide and pure carbon (soot or ash). Additionally, any combustion in atmospheric air, which is 78% nitrogen, will also create several forms of nitrogen oxides.

Types—Complete vs. Incomplete

In complete combustion, the reactant burns in oxygen, producing a limited number of products. When a hydrocarbon burns in oxygen, the reaction will only yield carbon dioxide and water. When elements are burned, the products are primarily the most common oxides. Carbon will yield carbon dioxide, nitrogen will yield nitrogen dioxide, sulfur will yield sulfur dioxide, and iron will yield iron(III) oxide.

Combustion is not necessarily favourable to the maximum degree of oxidation and it can be temperature-dependent. For example, sulfur trioxide is not produced quantitatively in combustion of sulfur. Nitrogen oxides start to form above 2,800 °F (1,540 °C) and more nitrogen oxides are produced at higher temperatures. Below this temperature, molecular nitrogen (N_2) is favoured. It is also a function of oxygen excess.

In most industrial applications and in fires, air is the source of oxygen (O_2). In air, each mole of oxygen is mixed with approximately 3.76 mole of nitrogen. Nitrogen does not take part in combustion, but at high temperatures, some nitrogen will be converted to NO_x, usually between 1% and 0.002% (2 ppm). Furthermore, when there is any incomplete combustion, some of carbon is converted to carbon monoxide. A more complete set of equations for combustion of methane in air is therefore:

$$CH_4 + 2\ O_2 \rightarrow CO_2 + 2\ H_2O$$

$$2\ CH_4 + 3\ O_2 \rightarrow 2\ CO + 4\ H_2O$$

$$N_2 + O_2 \rightarrow 2\ NO$$

$$N_2 + 2\ O_2 \rightarrow 2\ NO_2$$

Incomplete

Incomplete combustion will only occur when there is not enough oxygen to allow the fuel to react completely to produce carbon dioxide and water. It also happens when the combustion is quenched by a heat sink such as a solid surface or flame trap.

For most fuels, such as diesel oil, coal or wood, pyrolysis occurs before combustion. In incomplete combustion, products of pyrolysis remain unburnt and contaminate the smoke with noxious particulate matter and gases. Partially oxidised compounds are also a concern; partial oxidation of ethanol can produce harmful acetaldehyde, and carbon can produce toxic carbon monoxide.

The quality of combustion can be improved by design of combustion devices, such as burners and internal combustion engines. Further improvements are achievable by catalytic after-burning devices (such as catalytic converters) or by the simple partial return of the exhaust gases into the combustion process. Such devices are required by environmental legislation for cars in most countries, and may be necessary in large combustion devices, such as thermal power plants, to reach legal emission standards.

The degree of combustion can be measured and analysed, with test equipment. HVAC contractors, firemen and engineers use combustion analysers to test the efficiency of a burner during the combustion process. In addition, the efficiency of an internal combustion engine can be measured in this way, and some states and local municipalities are using combustion analysis to define and rate the efficiency of vehicles on the road today.

Smoldering

Smoldering is the slow, low-temperature, flameless form of combustion, sustained by the heat evolved when oxygen directly attacks the surface of a condensed-phase fuel. It is a typically incomplete combustion reaction. Solid materials that can sustain a smoldering reaction include coal, cellulose, wood, cotton, tobacco, peat, duff, humus, synthetic foams, charring polymers including polyurethane foam, and dust. Common examples of smoldering phenomena are the initiation of residential fires on upholstered furniture by weak heat sources (e.g., a cigarette, a short-circuited wire), and the persistent combustion of biomass behind the flaming front of wildfires

Rapid

Rapid combustion is a form of combustion, otherwise known as a fire, in which large amounts of heat and light energy are released, which often results in a flame. This is used in a form of machinery such as internal combustion engines and in thermobaric weapons. Sometimes, a large volume of gas is liberated in combustion besides the production of heat and light. The sudden evolution of large quantities of gas creates excessive pressure that produces a loud noise. Such a combustion is known as an explosion. Combustion need not involve oxygen; e.g., hydrogen burns in chlorine to form hydrogen chloride with the liberation of heat and light characteristic of combustion.

Turbulent

Combustion resulting in a turbulent flame is the most used for industrial application (e.g. gas turbines, gasoline engines, etc.) because the turbulence helps the mixing process between the fuel and oxidiser.

Microgravity

Combustion processes behave differently in a microgravity environment than in Earth-gravity conditions due to the lack of buoyancy. For example, a candle's flame takes the shape of a sphere. Microgravity combustion research contributes to understanding of spacecraft fire safety and diverse aspects of combustion physics.

Micro Combustion

Combustion processes which happen in very small volume are considered as micro combustion. Quenching distance plays a vital role in stabilising the flame in such combustion chambers.

Chemical Equation

Generally, the chemical equation for stoichiometric burning of hydrocarbon in oxygen is

$$C_xH_y + \left(x + \frac{y}{4}\right)O_2 \rightarrow xCO_2 + \left(\frac{y}{2}\right)H_2O$$

For example, the burning of propane is

$$C_3H_8 + 5O_2 \rightarrow 3CO_2 + 4H_2O$$

Generally, the chemical equation for stoichiometric incomplete combustion of hydrocarbon in oxygen is as follows:

$$(z)C_xH_y + \left(z\left(\frac{x}{2} + \frac{y}{4}\right)\right)O_2 \rightarrow z\,xCO + \left(\frac{z\,y}{2}\right)H_2O$$

For example, the incomplete combustion of propane is:

$$2C_3H_8 + 7O_2 \rightarrow 2C + 2CO + 8H_2O + 2CO_2$$

The simple word equation for the combustion of a hydrocarbon in oxygen is:

$$\text{Fuel} + \text{Oxygen} \rightarrow \text{Heat} + \text{Water} + \text{Carbon dioxide}$$

If the combustion takes place using air as the oxygen source, the nitrogen can be added to the equation, as and although it does not react, to show the composition of the flue gas:

$$C_xH_y + \left(x + \frac{y}{4}\right)O_2 + 3.76\left(x + \frac{y}{4}\right)N_2 \rightarrow xCO_2 + \left(\frac{y}{2}\right)H_2O + 3.$$

For example, the burning of propane is:

$$C_3H_8 + 5O_2 + 18.8N_2 \rightarrow 3CO_2 + 4H_2O + 18.8N_2$$

The simple word equation for this type of combustion is hydrocarbon in air:

$$\text{Fuel} + \text{Air} \rightarrow \text{Heat} + \text{Water} + \text{Carbon dioxide} + \text{Nitrogen}$$

Nitrogen may also oxidise when there is an excess of oxygen. The reaction is thermodynamically favoured only at high temperatures. Diesel engines are run with an excess of oxygen to combust small particles that tend to form with only a stoichiometric amount of oxygen, necessarily producing nitrogen oxide emissions. Both the United States and European Union are planning to impose limits to nitrogen oxide emissions, which necessitate the use of a special catalytic converter or treatment of the exhaust with urea.

Fuels—Liquid Fuels

Combustion of a liquid fuel in an oxidising atmosphere actually happens in the gas phase. It is the vapour that burns, not the liquid. Therefore, a liquid will normally catch fire only above a certain temperature: its flash point. The flash point of a liquid fuel is the lowest temperature at which it can form an ignitable mix with air.

It is also the minimum temperature at which there is enough evapourated fuel in the air to start combustion.

Solid Fuels

The act of combustion consists of three relatively distinct but overlapping phases:

- Preheating phase, when the unburned fuel is heated up to its flash point and then fire point. Flammable gases start being evolved in a process similar to dry distillation.
- Distillation phase or gaseous phase, when the mix of evolved flammable gases with oxygen is ignited. Energy is produced in the form of heat and light. Flames are often visible. Heat transfer from the combustion to the solid maintains the evolution of flammable vapours.
- Charcoal phase or solid phase, when the output of flammable gases from the material is too low for persistent presence of flame and the charred fuel does not burn rapidly and just glows and later only smoulders.

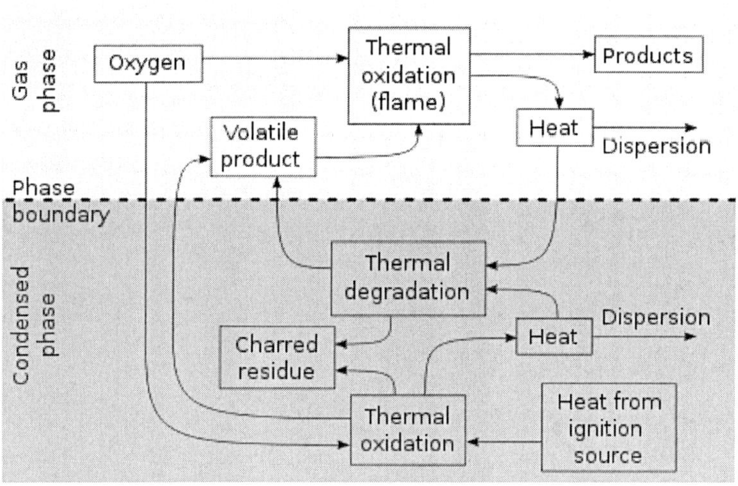

Figure 9: A general scheme of polymer combustion

Reaction Mechanism

Combustion in oxygen is a radical chain reaction where many distinct radical intermediates participate.

The high energy required for initiation is explained by the unusual structure of the dioxygen molecule. The lowest-energy configuration of the dioxygen molecule is a stable, relatively unreactive diradical in a triplet spin state. Bonding can be described with three bonding electron pairs and two antibonding electrons, whose spins are aligned, such that the molecule has nonzero total angular momentum. Most fuels, on the other hand, are in a singlet state, with paired spins and zero total angular momentum. Interaction between the two is quantum mechanically a "forbidden transition", i.e. possible with a very low probability. To initiate combustion, energy is required to force dioxygen into a spin-paired state, or singlet oxygen. This intermediate is extremely reactive. The energy is supplied as heat. The reaction produces heat, which keeps it going.

Combustion of hydrocarbons is thought to be initiated by hydrogen atom abstraction (not proton abstraction) from the fuel to oxygen, to give a hydroperoxide radical (HOO). This reacts further to give hydroperoxides, which break up to give hydroxyl radicals. There are a great variety of these processes that produce fuel radicals and oxidising radicals. Oxidising species include singlet oxygen, hydroxyl, monatomic oxygen, and hydroperoxyl. Such intermediates are short-lived and cannot be isolated. However, non-radical intermediates are stable and are produced in incomplete combustion. An example is acetaldehyde produced in the combustion of ethanol. An intermediate in the combustion of carbon and hydrocarbons, carbon monoxide, is of special importance because it is a poisonous gas, but also economically useful for the production of syngas.

Solid fuels also undergo a great number of pyrolysis reactions that give more easily oxidised, gaseous fuels. These reactions are endothermic and require constant energy input from the combustion reactions. A lack of oxygen or other poorly designed conditions result in these noxious and carcinogenic pyrolysis products being emitted as thick, black smoke.

Temperature

Assuming perfect combustion conditions, such as complete combustion under adiabatic conditions (i.e., no heat loss or gain), the adiabatic combustion temperature can be determined. The formula that yields this temperature is based on the first law of thermodynamics and takes note of the fact that the heat of combustion is used entirely for heating the fuel, the combustion air or oxygen, and the combustion product gases (commonly referred to as the *flue gas*).

In the case of fossil fuels burnt in air, the combustion temperature depends on all of the following:

- the heating value;

- the stoichiometric air to fuel ratio λ;
- the specific heat capacity of fuel and air;
- the air and fuel inlet temperatures.

The adiabatic combustion temperature (also known as the *adiabatic flame temperature*) increases for higher heating values and inlet air and fuel temperatures and for stoichiometric air ratios approaching one.

Most commonly, the adiabatic combustion temperatures for coals are around 2,200 °C (3,992 °F) (for inlet air and fuel at ambient temperatures and for λ = 1.0), around 2,150 °C (3,902 °F) for oil and 2,000 °C (3,632 °F) for natural gas.

In industrial fired heaters, power plant steam generators, and large gas-fired turbines, the more common way of expressing the usage of more than the stoichiometric combustion air is *percent excess combustion air*. For example, excess combustion air of 15 percent means that 15 percent more than the required stoichiometric air is being used.

Instabilities

Combustion instabilities are typically violent pressure oscillations in a combustion chamber. These pressure oscillations can be as high as 180 dB, and long term exposure to these cyclic pressure and thermal loads reduces the life of engine components. In rockets, such as the F1 used in the Saturn V program, instabilities led to massive damage of the combustion chamber and surrounding components. This problem was solved by re-designing the fuel injector. In liquid jet engines the droplet size and distribution can be used to attenuate the instabilities. Combustion instabilities are a major concern in ground-based gas turbine engines because of NOx emissions. The tendency is to run lean, an equivalence ratio less than 1, to reduce the combustion temperature and thus reduce the NOx emissions; however, running the combustion lean makes it very susceptible to combustion instabilities.

The Rayleigh Criterion is the basis for analysis of thermoacoustic combustion instabilities and is evaluated using the Rayleigh Index over one cycle of instability:

$$G(x) = \frac{1}{T} \int_T q'(x,t) p'(x,t) dt$$

where q' is the heat release rate perturbation and p' is the pressure fluctuation. When the heat release oscillations are in phase with the pressure oscillations, the Rayleigh Index is positive and the magnitude of the thermo acoustic instability increases. On the other hand, if the Rayleigh Index is negative, then thermoacoustic damping occurs. The Rayleigh Criterion implies that a thermoacoustic instability can be optimally controlled by having heat release oscillations 180 degrees out of phase with pressure oscillations at the same frequency. This minimises the Rayleigh Index.

Rate of Combustion

Rate of combustion is the amount of mass of a material that goes through combustion over an amount of time. It can be expressed in g/s or kg/s.

Micro Combustion

Micro Combustion is the sequence of exothermic chemical reaction between a fuel and an oxidant accompanied by the production of heat and conversion of chemical species at micro level. The release of heat can result in the production of light in the form of either glowing or a flame. Fuels of interest often include organic compounds (especially hydrocarbons) in the gas, liquid or solid phase.

The development of miniaturised products such as microrobots, notebook computers, micro-aerial vehicles and other small scale devices is becoming increasingly important in our daily life. There is a growing interest in developing small scale combustor to power these micro devices due to their inherent advantages of higher energy-density, higher heat and mass transfer coefficients and smaller recharge times compared to electrochemical batteries. Energy-density of hydrocarbon fuels is 20-50 times higher than the most advanced Li-ion concept based electrochemical batteries. Concept of micro heat engine was proposed by Epstein and Senturia in 1997.

Since then, substantial amount of work has been done towards the development and application of such small scale devices to generate power through the combustion of hydrocarbon fuels. Micro combustors are an attractive alternate to batteries as they have large surface area to volume ratio, due to which, significant amount of heat is transferred through the walls which leads to came quenching. However, the increased rate of heat transfer through solid walls is advantageous in the case of steam reformers used for Hydrogen production.

B. Khandelwal et al.. have experimentally studied the flame stability limits and other characteristics in a two staged micro combustor. They found out that staged combustor leads to higher flame stability limits, in addition to that they also offer higher temperature profiles which would be helpful in utilising the heat produced by combustion. Maruta et al. have experimentally studied the came propagation characteristics of premixed methane air mixtures in a 2.0 mm diametre straight quartz channel with a positive wall temperature gradient along the how direction. This was a simple one-dimensional configuration to study came stabilisation characteristics in microchannels. Other researchers have studied the came stabilisation behaviour and combustion performance in Swiss roll combustor , micro-gas turbine engines , micro thermo photovoltaic system , free piston knock engine , micro-tube combustor , radial channel combustors and various other concepts of micro combustors .

Major problem of micro combustion is high surface to volume ratio. As the surface to volume ratio increases heat loss to walls of combustor increases which leads to flame quenching.

Quenching

In materials science, quenching is the rapid cooling of a workpiece to obtain certain material properties. It prevents low-temperature processes, such as phase transformations, from occurring by only providing a narrow window of time in which the reaction is both thermodynamically favourable and kinetically accessible. For instance, it can reduce crystallinity and thereby increase toughness of both alloys and plastics (produced through polymerisation).

In metallurgy, it is most commonly used to harden steel by introducing martensite, in which case the steel must be rapidly cooled through its eutectoid point, the temperature at which austenite becomes unstable. In steel alloyed with metals such as nickel and manganese, the eutectoid temperature becomes much lower, but the kinetic barriers to phase transformation remain the same. This allows quenching to start at a lower temperature, making the process much easier. High speed steel also has added tungsten, which serves to raise kinetic barriers and give the illusion that the material has been cooled more rapidly than it really has. Even cooling such alloys slowly in air has most of the desired effects of quenching.

Extremely rapid cooling can prevent the formation of all crystal structure, resulting in amorphous metal or “metallic glass”.

Quench Hardening

Quench hardening is a mechanical process in which steel and cast iron alloys are strengthened and hardened. These metals consist of ferrous metals and alloys. This is done by heating the material to a certain temperature, dependent upon material, and then rapidly cooling the material. This produces a harder material by either surface hardening or through-hardening varying on the rate at which the material is cooled. The material is then often tempered to reduce the brittleness that may increase from the quench hardening process. Items that may be quenched include gears, shafts, and wear blocks.

Process

Quenching metals is a progression; the first step is soaking the metal, i.e. heating it to the required temperature. Soaking can be done by air (air furnace), or a bath. The soaking time in air furnaces should be 1 to 2 minutes for each millimetre of cross-section. For a bath the time can range a little higher. The recommended time allotment in salt or lead baths is 0 to 6 minutes. Uneven heating or overheating should be avoided at all cost. Most materials are heated from anywhere to 815 to 900 °C (1,500 to 1,650 °F).

The next item on the progression list is the cooling of the part. Water is one of the most efficient quenching media where maximum hardness is acquired, but there is a small chance that it may cause distortion and tiny cracking. When hardness can be sacrificed, whale, cottonseed and mineral oils are used. These often tend to oxidise and form a sludge, which consequently lowers the efficiency. The quenching velocity (cooling rate) of oil is much less than water. Intermediate rates between water and oil can be obtained with water containing 10-30% Ucon, a substance with an inverse solubility which therefore deposits on the object to slow the rate of cooling.

To minimise distortion, long cylindrical workpieces are quenched vertically; flat workpieces are quenched on edge; and thick sections should enter the bath first. To prevent steam bubbles the bath is agitated.

Effect of Quench Hardening

Before the material is hardened, the microstructure of the material is a pearlite grain structure that is uniform and laminar. Pearlite is a mixture of ferrite and cementite formed when steel or cast iron are manufactured and cooled at a slow rate. After quench hardening, the microstructure of the material form into martensite as a fine, needle-like grain structure.

Before using this technique it is essential to look up the rate constants for the quenching of the excited states of metal ions.

Equipment

There are three types of furnaces that are commonly used in quench hardening: salt bath furnace, continuous furnace, and box furnace. Each is used depending on what other processes or types of quench hardening are being done on the different materials.

Quenching Media

When quenching, there are numerous types of media. Some of the more common include: air, brine (salt water), oil and water. These media are used to increase the severity of the quench.

Chapter 8

Air Standard Cycles

Theoretical Analysis

The accurate analysis of the various processes taking place in an internal combustion engine is a very complex problem. If these processes were to be analysed experimentally, the analysis would be very realistic no doubt. It would also be quite accurate if the tests are carried out correctly and systematically, but it would be time consuming. If a detailed analysis has to be carried out involving changes in operating parameters, the cost of such an analysis would be quite high, even prohibitive. An obvious solution would be to look for a quicker and less expensive way of studying the engine performance characteristics. A theoretical analysis is the obvious answer.

A theoretical analysis, as the name suggests, involves analysing the engine performance without actually building and physically testing an engine. It involves *simulating* an engine operation with the help of thermodynamics so as to formulate mathematical expressions which can then be solved in order to obtain the relevant information. The method of solution will depend upon the complexity of the formulation of the mathematical expressions which in turn will depend upon the assumptions that have been introduced in order to analyse the processes in the engine. The more the assumptions, the simpler will be the mathematical expressions and the easier the calculations, but the lesser will be the accuracy of the final results.

The simplest theoretical analysis involves the use of the air standard cycle, which has the largest number of simplifying assumptions.

A Thermodynamic Cycle

In some practical applications, notably steam power and refrigeration, a thermodynamic cycle can be identified. A thermodynamic cycle occurs when the working fluid of a system experiences a number of processes that eventually return the fluid to its initial state.

In steam power plants, water is pumped (for which work W_P is required) into a boiler and evapourated into steam while heat Q_A is supplied at a high temperature.

The steam flows through a turbine doing work W_T and then passes into a condenser where it is condensed into water with consequent rejection of heat Q_R to the atmosphere. Since the water is returned to its initial state, the net change in energy is zero, assuming no loss of water through leakage or evapouration.

An energy equation pertaining only to the system can be derived. Considering a system with one entering and one leaving flow stream for the time period t_1 to t_2

$$\Delta Q - \Delta W + \Delta E_{f_{in}} - \Delta E_{f_{out}} = \Delta E_{system} \tag{1}$$

ΔQ is the heat transfer across the boundary,

+ve for heat *added to* the system and

–ve for heat *taken from* the system.

W is the work transfer across the boundary,

+ve for work *done by* the system and

-ve for work *added to* the system $\Delta E_{f_{in}}$

is the energy of all forms *carried* by the fluid across the boundary *into* the system

is the energy of all forms *carried* by the fluid across the boundary *out* of system

ΔE_{system} is the energy of all forms *stored* within the system,

+ve for energy *increase*

-ve for energy *decrease*

In the case of the steam power system described above

$$Q_A + Q_R = \sum Q = \sum W = W_T + W_P \tag{2}$$

All thermodynamic cycles have a heat rejection process as an invariable characteristic and the net work done is always less than the heat supplied, although, as shown in Eq. 2, it is equal to the sum of heat added and the heat rejected (Q_R is a negative number).

The thermal efficiency of a cycle, η_{th}, is defined as the fraction of heat supplied to a thermodynamic cycle that is converted to work, that is

$$\eta_{th} = \frac{\sum W}{Q_A}$$

$$= \frac{Q_A + Q_R}{Q_A} \tag{3}$$

This efficiency is sometimes confused with the enthalpy efficiency, η_e, or the fuel conversion efficiency, η_f

$$\eta_e = \frac{\sum W}{m_f Q_c} \tag{4}$$

This definition applies to combustion engines which have as a source of energy the chemical energy residing in a fuel used in the engine.

Any device that operated in a thermodynamic cycle, absorbs thermal energy from a source, rejects a part of it to a sink and presents the difference between the energy absorbed and energy rejected *as work to the surroundings* is called a heat engine.

A heat engine is, thus, a device that produces work. In order to achieve this purpose, the heat engine uses a certain working medium which undergoes the following processes:

1. A compression process where the working medium absorbs energy as work.
2. A heat addition process where the working medium absorbs energy as heat from a source.
3. An expansion process where the working medium transfers energy as work to the surroundings.
4. A heat rejection process where the working medium rejects energy as heat to a sink.

If the working medium does not undergo any change of phase during its passage through the cycle, the heat engine is said to operate in a non-phase change cycle. A phase change cycle is one in which the working medium undergoes changes of phase. The air standard cycles, using air as the working medium are examples of non-phase change cycles while the steam and vapour compression refrigeration cycles are examples of phase change cycles.

Air Standard Cycles

The air standard cycle is a cycle followed by a heat engine which uses air as the working medium. Since the air standard analysis is the simplest and most idealistic, such cycles are also called *ideal cycles* and the engine running on such cycles are called *ideal engines.*

In order that the analysis is made as simple as possible, certain assumptions have to be made. These assumptions result in an analysis that is far from correct for most actual combustion engine processes, but the analysis is of considerable value for indicating the upper limit of performance. The analysis is also a simple means for indicating the relative effects of principal variables of the cycle and the relative size of the apparatus.

Assumptions

1. The working medium is a perfect gas with constant specific heats and molecular weight corresponding to values at room temperature.
2. No chemical reactions occur during the cycle. The heat addition and heat rejection processes are merely heat transfer processes.
3. The processes are reversible.
4. Losses by heat transfer from the apparatus to the atmosphere are assumed to be zero in this analysis.
5. The working medium at the end of the process (cycle) is unchanged and is at the same condition as at the beginning of the process (cycle).

In The selecting an idealised process one is always faced with the fact that the simpler the assumptions, the easier the analysis, but the farther the result from reality. The air cycle has the advantage of being based on a few simple assumptions and of lending itself to rapid and easy mathematical handling without recourse to thermodynamic charts or tables or complicated calculations. On the other hand, there is always the danger of losing sight of its limitations and of trying to employ it beyond its real usefulness.

Equivalent Air Cycle

A particular air cycle is usually taken to represent an approximation of some real set of processes which the user has in mind. Generally speaking, the air cycle representing a given real cycle is called an *equivalent air cycle.* The equivalent cycle has, in general, the following characteristics in common with the real cycle which it approximates:

1. A similar sequence of processes.
2. Same ratio of maximum to minimum volume for reciprocating engines or maximum to minimum pressure for gas turbine engines.
3. The same pressure and temperature at a given reference point.
4. An appropriate value of heat addition per unit mass of air.

The Carnot Cycle

This cycle was proposed by Sadi Carnot in 1824 and has the highest possible efficiency for any cycle.

Assuming that the charge is introduced into the engine at point 1, it undergoes isentropic compression from 1 to 2. The temperature of the charge rises from T_{min} to T_{max}.

At point 2, heat is added isothermally. This causes the air to expand, forcing the piston forward, thus doing work on the piston. At point 3, the source of heat is removed and the air now expands isentropically to point 4, reducing the temperature to T_{min} in the process.

At point 4, a cold body is applied to the end of the cylinder and the piston reverses, thus compressing the air isothermally; heat is rejected to the cold body. At point 1, the cold body is removed and the charge is compressed isentropically till it reaches a temperature T_{max} once again.

Thus, the heat addition and rejection processes are isothermal while the compression and expansion processes are isentropic.

From thermodynamics, per unit mass of charge

$$\text{Heat supplied from point 1 to 2} = p_2 v_2 \ln\frac{v_3}{v_2} \tag{5}$$

$$\text{Heat rejected from point 4 to 1} = p_4 v_4 \ln\frac{v_1}{v_4} \tag{6}$$

$$\text{Now } p_2 v_2 = RT_{max} \tag{7}$$

$$\text{And } p_4 v_4 = RT_{min} \tag{8}$$

Since Work done, per unit mass of charge, W = heat supplied – heat rejected

$$W = RT_{max} \ln \frac{v_3}{v_2} - RT_{min} \ln \frac{v_1}{v_4}$$

$$= R \ln(r)(T_{max} - T_{min}) \quad (9)$$

We have assumed that the compression and expansion ratios are equal, that is

$$\frac{v_3}{v_2} = \frac{v_1}{v_4} \quad (10)$$

Heat supplied $Q_s = R\ T_{max} \ln (r)$

Hence, the thermal efficiency of the cycle is given by

$$\eta_{th} = \frac{R \ln(r)(T_{max} - T_{min})}{R \ln(r) T_{max}}$$

$$= \frac{T_{max} - T_{min}}{T_{max}} \quad (11)$$

From Eq. 11 it is seen that the thermal efficiency of the Carnot cycle is only a function of the maximum and minimum temperatures of the cycle. The efficiency will increase if the minimum temperature (or the temperature at which the heat is rejected) is as low as possible. According to this equation, the efficiency will be equal to 1 if the minimum temperature is zero, which happens to be the absolute zero temperature in the thermodynamic scale.

This equation also indicates that for optimum (Carnot) efficiency, the cycle (and hence the heat engine) must operate between the limits of the highest and lowest possible temperatures. In other words, the engine should take in all the heat at as high a temperature as possible and should reject the heat at as low a temperature as possible. For the first condition to be achieved, combustion (as applicable for a real engine using fuel to provide heat) should begin at the highest possible temperature, for then the irreversibility of the chemical reaction would be reduced. Moreover, in the cycle, the expansion should proceed to the lowest possible temperature in order to obtain the maximum amount of work. These conditions are the aims of all designers of modern heat engines. The conditions of heat rejection are governed, in practice, by the temperature of the atmosphere.

It is impossible to construct an engine which will work on the Carnot cycle. In such an engine, it would be necessary for the piston to move very slowly during the first part of the forward stroke so that it can follow an isothermal process. During the remainder of the forward stroke, the piston would need to move very quickly as it has to follow an isentropic process. This variation in the speed of the piston cannot be achieved in practice. Also, a very long piston stroke would produce only a small amount of work most of which would be absorbed by the friction of the moving parts of the engine.

Since the efficiency of the cycle, as given by Eq. 11, is dependent only on the maximum and minimum temperatures, it does not depend on the working medium. It is thus independent of the properties of the working medium.

Piston Engine Air Standard Cycles

The cycles described here are air standard cycles applicable to piston engines. Engines bases on these cycles have been built and many of the engines are still in use.

The Lenoir Cycle

The Lenoir cycle is of interest because combustion (or heat addition) occurs without compression of the charge.

According to the cycle, the piston is at the top dead centre, point 1, when the charge is ignited (or heat is added). The process is at constant volume so the pressure rises to point 2. From 2 to 3, expansion takes place and from 3 to 1 heat is rejected at constant pressure.

Heat supplied, $Q_s = c_v(T_2 - T_1)$ (12)

Heat rejected, $Q_r = c_p(T_3 - T_1)$ (13)

Since $W = Q_s - Q_r$ (14)

$W = c_v(T_2 - T_1) - c_p(T_2 - T_1)$ (15)

$$\text{Thus} \quad \eta_{th} = 1 - \frac{c_p(T_3 - T_1)}{c_v(T_2 - T_1)} \tag{16}$$

$$= 1 - \frac{\gamma\left(\frac{T_3}{T_1} - 1\right)}{\left(\frac{T_2}{T_1} - 1\right)} \tag{17}$$

$$\text{Since} \quad \frac{T_2}{T_1} = \frac{p_2}{p_1}, \quad \frac{T_3}{T_1} = \frac{V_3}{V_1}$$

and $p_2 V_2^\gamma = p_3 V_3^\gamma$ so $\dfrac{p_2}{p_3} = \left(\dfrac{V_3}{V_2}\right)^\gamma = \left(\dfrac{V_3}{V_1}\right)^\gamma = \dfrac{p_2}{p_1} = \dfrac{T_2}{T_1}$

$$\therefore \quad \eta_{th} = 1 - \frac{\gamma\left(\dfrac{V_3}{V_1} - 1\right)}{\left[\left(\dfrac{V_3}{V_1}\right)^\gamma - 1\right]}$$

$$= 1 - \frac{\gamma(r_e - 1)}{(r_e^\gamma - 1)} \qquad (18)$$

Here, $r_e = V_3/V_1$, the volumetric expansion ratio. Equation 18 indicates that the thermal efficiency of the Lenoir cycle depends primarily on the expansion ratio and the ratio of specific heats.

The intermittent-flow engine which powered the German V-1 buzz-bomb in 1942 during World War II operated on a modified Lenoir cycle. A few engines running on the Lenoir cycle were built in the late 19th century till the early 20th century.

The Otto Cycle

The Otto cycle, which was first proposed by a Frenchman, Beau de Rochas in 1862, was first used on an engine built by a German, Nicholas A. Otto, in 1876. The cycle is also called a *constant volume* or *explosion* cycle. This is the equivalent air cycle for reciprocating piston engines using spark ignition.

At the start of the cycle, the cylinder contains a mass M of air at the pressure and volume indicated at point 1. The piston is at its lowest position.

It moves upward and the gas is compressed isentropically to point 2. At this point, heat is added at constant volume which raises the pressure to point 3. The high pressure charge now expands isentropically, pushing the piston down on its expansion stroke to point 4 where the charge rejects heat at constant volume to the initial state, point 1.

The isothermal heat addition and rejection of the Carnot cycle are replaced by the constant volume processes which are, theoretically more plausible, although in practice, even these processes are not practicable.

The heat supplied, Q_s, per unit mass of charge, is given by

$c_v(T_3 - T_2)$

the heat rejected, Q_r per unit mass of charge is given by

$c_v(T_4 - T_1)$

and the thermal efficiency is given by

$$\eta_{th} = 1 - \frac{(T_4 - T_1)}{(T_3 - T_2)}$$

$$= 1 - \frac{T_1}{T_2}\left\{\frac{\left(\frac{T_4}{T_1} - 1\right)}{\left(\frac{T_3}{T_2} - 1\right)}\right\} \quad (19)$$

Now $\frac{T_1}{T_2} = \left(\frac{V_2}{V_1}\right)^{\gamma - 1} = \left(\frac{V_3}{V_4}\right)^{\gamma - 1} = \frac{T_4}{T_3}$

And since $\frac{T_1}{T_2} = \frac{T_4}{T_3}$ *we have* $\frac{T_4}{T_1} = \frac{T_3}{T_2}$

Hence, substituting in Eq. 19, we get, assuming that r is the compression ratio V_1/V_2

$$\eta_{th} = 1 - \frac{T_1}{T_2}$$

$$= 1 - \left(\frac{V_2}{V_1}\right)^{\gamma - 1}$$

$$= 1 - \frac{1}{r^{\gamma - 1}} \quad (20)$$

In a true thermodynamic cycle, the term *expansion ratio* and *compression ratio* are synonymous. However, in a real engine, these two ratios need not be equal because of the valve timing and therefore the term *expansion ratio* is preferred sometimes.

Equation 20 shows that the thermal efficiency of the theoretical Otto cycle increases with increase in compression ratio and specific heat ratio but is independent of the heat added (independent of load) and initial conditions of pressure, volume and temperature.

It is seen that the increase in efficiency is significant at lower compression ratios. This is also seen in Table 1 given below.

R	*h*
1	0
2	0.242
3	0.356
4	0.426
5	0.475
6	0.512
7	0.541
8	0.565
9	0.585
10	0.602
16	0.67
20	0.698
50	0.791

From the table it is seen that if:

CR is increased from 2 to 4, efficiency increase is 76%

CR is increased from 4 to 8, efficiency increase is only 32.6%

CR is increased from 8 to 16, efficiency increase is only 18.6%

Mean Effective Pressure

It is seen that the air standard efficiency of the Otto cycle depends only on the compression ratio. However, the pressures and temperatures at the various points in the cycle and the net work done, all depend upon the initial pressure and temperature and the heat input from point 2 to point 3, besides the compression ratio.

A quantity of special interest in reciprocating engine analysis is the mean effective pressure.

Mathematically, it is the net work done on the piston, W, divided by the piston *displacement* volume, $V_1 - V_2$. This quantity has the units of pressure. Physically, it is that constant pressure which, if exerted on the piston for the whole outward stroke, would yield work equal to the work of the cycle. It is given by

$$mep = \frac{W}{V_1 - V_2}$$

$$= \frac{\eta Q_{2-3}}{V_1 - V_2} \quad (21)$$

where Q_{2-3} is the heat added from points 2 to 3.

Now

$$V_1 - V_2 = V_1\left(1 - \frac{V_2}{V_1}\right)$$

$$= V_1\left(1 - \frac{1}{r}\right) \quad (22)$$

Here r is the compression ratio, V_1/V_2

From the equation of state:

$$V_1 = M\frac{R_0}{m}\frac{T_1}{p_1} \quad (23)$$

R_0 is the universal gas constant

Substituting for V_1 from Eq. 3 in Eq. 2 and then substituting for $V_1 - V_2$ in Eq. 1 we get

$$mep = \eta\frac{Q_{2-3}\dfrac{p_1 m}{MR_0 T_1}}{1 - \dfrac{1}{r}} \quad (24A)$$

The quantity Q_{2-3}/M is the heat added between points 2 and 3 *per unit mass* of air (M is the mass of air and m is the molecular weight of air); and is denoted by Q', thus

$$mep = \eta\frac{Q'\dfrac{p_1 m}{R_0 T_1}}{1 - \dfrac{1}{r}} \quad (24B)$$

We can non-dimensionalise the mep by dividing it by p_1 so that we can obtain the following equation

$$\frac{mep}{p_1} = \eta\left[\frac{1}{1 - \dfrac{1}{r}}\right]\left[\frac{Q' m}{R_0 T_1}\right] \quad (25)$$

Since $\frac{R_0}{m} = c_v(\gamma - 1)$, we can substitute it in Eq. 25 to get

$$\frac{mep}{p_1} = \eta \frac{Q'}{c_v T_1} \frac{1}{\left[1 - \frac{1}{r}\right][\gamma - 1]} \tag{26}$$

The dimensionless quantity mep/p_1 is a function of the heat added, initial temperature, compression ratio and the properties of air, namely, c_v and γ. We see that the mean effective pressure is directly proportional to the heat added and inversely proportional to the initial (or ambient) temperature.

We can substitute the value of η from Eq. 20 in Eq. 26 and obtain the value of mep/p_1 for the Otto cycle in terms of the compression ratio and heat added.

In terms of the pressure ratio, p_3/p_2 denoted by r_p we could obtain the value of mep/p_1 as follows:

$$\frac{mep}{p_1} = \frac{r(r_p - 1)(r^{\gamma-1} - 1)}{(r - 1)(\gamma - 1)} \tag{27}$$

We can obtain a value of r_p in terms of Q' as follows:

$$r_p - \frac{Q'}{c_v T_1 r^{\gamma-1}} + 1 \tag{28}$$

Another parameter, which is of importance, is the quantity mep/p_3. This can be obtained from the following expression:

$$\frac{mep}{p_3} = \frac{mep}{p_1} \frac{1}{r^{\gamma}} \frac{1}{\frac{Q'}{c_v T_1 r^{\gamma-1}} + 1} \tag{29}$$

Choice of Q'

We have said that

$$Q' = \frac{Q_{2-3}}{M} \tag{30}$$

M is the mass of charge (air) per cycle, kg.

Now, in an actual engine

$$Q_{2-3} = M_f Q_c$$

$$= F M_a Q_c \ in\ kJ / cycle \tag{31}$$

M_f is the mass of fuel supplied per cycle, kg

Q_c is the heating value of the fuel, kJ/kg

M_a is the mass of air taken in per cycle

F is the fuel air ratio = M_f/M_a

Substituting for Eq. (B) in Eq. (A) we get

$$Q' = \frac{FM_a Q_c}{M} \tag{32}$$

$$Now \quad \frac{M_a}{M} \approx \frac{V_1 - V_2}{V_1}$$

$$And \quad \frac{V_1 - V_2}{V_1} = 1 - \frac{1}{r} \tag{33}$$

So, substituting for M_a/M from Eq. (33) in Eq. (32) we get

$$Q' = FQ_c\left(1 - \frac{1}{r}\right) \tag{34}$$

For isooctane, FQ_c at stoichiometric conditions is equal to 2975 kJ/kg, thus

$$Q' = 2975(r - 1)/r \tag{35}$$

At an ambient temperature, T_1 of 300K and c_v for air is assumed to be 0.718 kJ/kgK, we get a value of $Q'/c_vT_1 = 13.8(r - 1)/r$.

Under fuel rich conditions, $\varphi = 1.2$, $Q'/c_vT_1 = 16.6(r - 1)/r$.

Under fuel lean conditions, $\varphi = 0.8$, $Q'/c_vT_1 = 11.1(r - 1)/r$

The Diesel Cycle

This cycle, proposed by a German engineer, Dr. Rudolph Diesel to describe the processes of his engine, is also called the constant pressure cycle.

This is believed to be the equivalent air cycle for the reciprocating slow speed compression ignition engine.

The cycle has processes which are the same as that of the Otto cycle except that the heat is added at constant pressure.

The heat supplied, Q_s is given by

$$c_p(T_3 - T_2)$$

whereas the heat rejected, Q_r is given by

$$c_v(T_4 - T_1)$$

and the thermal efficiency is given by

$$\eta_{th} = 1 - \frac{c_v\left(T_4 - T_1\right)}{c_p\left(T_3 - T_2\right)}$$

$$= 1 - \frac{1}{\gamma}\left\{\frac{T_1\left(\dfrac{T_4}{T_1} - 1\right)}{T_2\left(\dfrac{T_3}{T_2} - 1\right)}\right\} \tag{36}$$

From the T-s diagram, the difference in enthalpy between points 2 and 3 is the same as that between 4 and 1, thus

$$\Delta s_{2-3} = \Delta s_{4-1}$$

$$\therefore\ c_v \ln\left(\frac{T_4}{T_1}\right) = c_p \ln\left(\frac{T_3}{T_2}\right)$$

$$\therefore\ \ln\left(\frac{T_4}{T_1}\right) = \gamma \ln\left(\frac{T_3}{T_2}\right)$$

$$\therefore\ \frac{T_4}{T_1} = \left(\frac{T_3}{T_2}\right)^{\gamma} \text{ and } \frac{T_1}{T_2} = \left(\frac{V_2}{V_1}\right)^{\gamma-1} = \frac{1}{r^{\gamma-1}}$$

Substituting in eq. 36, we get

$$\eta_{th} = 1 - \frac{1}{\gamma}\left(\frac{1}{r}\right)^{\gamma-1}\left[\frac{\left(\dfrac{T_3}{T_2}\right)^{\gamma} - 1}{\dfrac{T_3}{T_2} - 1}\right] \tag{37}$$

Now $\dfrac{T_3}{T_2} = \dfrac{V_3}{V_2} = r_c = cut-off\ ratio$

$$\eta = 1 - \frac{1}{r^{\gamma-1}}\left[\frac{r_c^{\gamma} - 1}{\gamma\left(r_c - 1\right)}\right] \tag{38}$$

When Eq. 38 is compared with Eq. 20, it is seen that the expressions are similar except for the term in the parentheses for the Diesel cycle. It can be shown that this term is always greater than unity.

Now $r_c = \frac{V_3}{V_2} = \frac{V_3}{V_4} \Big/ \frac{V_2}{V_1} = \frac{r}{r_e}$ where r is the compression ratio and r_e is the expansion ratio

Thus, the thermal efficiency of the Diesel cycle can be written as

$$\eta = 1 - \frac{1}{r^{\gamma-1}} \left[\frac{\left(\frac{r}{r_e}\right)^{\gamma} - 1}{\gamma\left(\frac{r}{r_e} - 1\right)} \right] \tag{39}$$

Let $r_e = r - \Delta$ since r is greater than r_e. Here, Δ is a small quantity. We therefore have

$$\frac{r}{r_e} = \frac{r}{r-\Delta} = \frac{r}{r\left(1 - \frac{\Delta}{r}\right)} = \left(1 - \frac{\Delta}{r}\right)^{-1}$$

We can expand the last term binomially so that

$$\left(1 - \frac{\Delta}{r}\right)^{-1} = 1 + \frac{\Delta}{r} + \frac{\Delta^2}{r^2} + \frac{\Delta^3}{r^3} + \cdots$$

Also $$\left(\frac{r}{r_e}\right)^{\gamma} = \frac{r^{\gamma}}{(r-\Delta)^{\gamma}} = \frac{r^{\gamma}}{r^{\gamma}\left(1 - \frac{\Delta}{r}\right)^{\gamma}} = \left(1 - \frac{\Delta}{r}\right)^{-\gamma}$$

We can expand the last term binomially so that

$$\left(1 - \frac{\Delta}{r}\right)^{-\gamma} = 1 + \gamma\frac{\Delta}{r} + \frac{\gamma(\gamma+1)}{2!}\frac{\Delta^2}{r^2} + \frac{\gamma(\gamma+1)(\gamma+2)}{3!}\frac{\Delta^3}{r^3} + \cdot$$

Substituting in Eq. 39, we get

$$\eta = 1 - \frac{1}{r^{\gamma-1}} \left[\frac{\frac{\Delta}{r} + \frac{(\gamma+1)}{2!}\frac{\Delta^2}{r^2} + \frac{(\gamma+1)(\gamma+2)}{3!}\frac{\Delta^3}{r^3} + \cdots}{\frac{\Delta}{r} + \frac{\Delta^2}{r^2} + \frac{\Delta^3}{r^3} + \cdots} \right] \tag{40}$$

Since the coefficients of $\frac{\Delta}{r}, \frac{\Delta^2}{r^r}, \frac{\Delta^3}{r^3}$, etc are greater than unity, the quantity in the brackets in Eq. 40 will be greater than unity. Hence, for the Diesel cycle, we subtract $\frac{1}{r^{\gamma-1}}$ times a quantity greater than unity from one, hence for the same r, the Otto cycle efficiency is greater than that for a Diesel cycle.

If $\frac{\Delta}{r}$ is small, the square, cube, etc of this quantity becomes progressively smaller, so the thermal efficiency of the Diesel cycle will tend towards that of the Otto cycle.

From the foregoing we can see the importance of cutting off the fuel supply early in the forward stroke, a condition which, because of the short time available and the high pressures involved, introduces practical difficulties with high speed engines and necessitates very rigid fuel injection gear.

In practice, the diesel engine shows a better efficiency than the Otto cycle engine because the compression of air alone in the former allows a greater compression ratio to be employed. With a mixture of fuel and air, as in practical Otto cycle engines, the maximum temperature developed by compression must not exceed the self ignition temperature of the mixture; hence a definite limit is imposed on the maximum value of the compression ratio.

Thus Otto cycle engines have compression ratios in the range of 7 to 12 while diesel cycle engines have compression ratios in the range of 16 to 22.

We can obtain a value of r_c for a Diesel cycle in terms of Q' as follows:

$$r_c = \frac{Q'}{c_p T_1 r^{\gamma-1}} + 1 \tag{41}$$

We can substitute the value of η from Eq. 38 in Eq. 26, reproduced below and obtain the value of mep/p_1 for the Diesel cycle.

$$\frac{mep}{p_1} = \eta \frac{Q'}{c_v T_1} \frac{1}{\left[1 - \frac{1}{r}\right][\gamma - 1]}$$

In terms of the cut-off ratio, we can obtain another expression for mep/p_1 as follows:

$$\frac{mep}{p_1}=\frac{\gamma r^{\gamma}\left(r_c-1\right)-r\left(r_c^{\gamma}-1\right)}{\left(r-1\right)\left(\gamma-1\right)} \tag{42}$$

For the Diesel cycle, the expression for mep/p_3 is as follows:

$$\frac{mep}{p_3}=\frac{mep}{p_1}\left(\frac{1}{r^{\gamma}}\right) \tag{43}$$

Modern high speed diesel engines do not follow the Diesel cycle. The process of heat addition is partly at constant volume and partly at constant pressure. This brings us to the dual cycle.

The Dual Cycle

An important characteristic of real cycles is the ratio of the mean effective pressure to the maximum pressure, since the mean effective pressure represents the useful (average) pressure acting on the piston while the maximum pressure represents the pressure which chiefly affects the strength required of the engine structure. It is seen that the quantity mep/p_3 falls off rapidly as the compression ratio increases, which means that for a given mean effective pressure the maximum pressure rises rapidly as the compression ratio increases.

For example, for a mean effective pressure of 7 bar and Q'/c_vT_1 of 12, the maximum pressure at a compression ratio of 5 is 28 bar whereas at a compression ratio of 10, it rises to about 52 bar. Real cycles follow the same trend and it becomes a practical necessity to limit the maximum pressure when high compression ratios are used, as in diesel engines. This also indicates that diesel engines will have to be stronger (and hence heavier) because it has to withstand higher peak pressures.

Constant pressure heat addition achieves rather low peak pressures unless the compression ratio is quite high. In a real diesel engine, in order that combustion takes place at constant pressure, fuel has to be injected very late in the compression stroke (practically at the top dead centre).

But in order to increase the efficiency of the cycle, the fuel supply must be cut off early in the expansion stroke, both to give sufficient time for the fuel to burn and thereby increase combustion efficiency and reduce after burning but also reduce emissions. Such situations can be achieved if the engine was a slow speed type so that the piston would move sufficiently slowly for combustion to take place despite the late injection of the fuel. For modern high speed compression ignition engines it is not possible to achieve constant pressure combustion. Fuel is injected somewhat earlier in the compression stroke and has to go through the various stages of combustion. Thus it is seen that combustion is nearly at constant volume (like in a spark ignition engine). But the peak pressure is limited because of strength considerations so the rest of the heat addition is believed to take place at constant pressure in a cycle. This has led to the formulation of the dual combustion cycle. In this cycle, for high compression ratios, the peak pressure is not allowed to increase beyond a certain limit and to account for the total addition, the rest of the heat is assumed to be added at constant pressure. Hence the name *limited pressure cycle.*

The cycle is the equivalent air cycle for reciprocating high speed compression ignition engines. The P-V and T-s diagrams are shown in figures. In the cycle, compression and expansion processes are isentropic; heat addition is partly at constant volume and partly at constant pressure while heat rejection is at constant volume as in the case of the Otto and Diesel cycles.

The heat supplied, Q_s per unit mass of charge is given by

$$c_{v(}T_3 - T_{2)} + c_{p(}T_{3'} - T_{2)}$$

whereas the heat rejected, Q_r per unit mass of charge is given by

$$c_{v(}T_4 - T_{1)}$$

and the thermal efficiency is given by

$$\eta_{th} = 1 - \frac{c_v\left(T_4 - T_1\right)}{c_v\left(T_3 - T_2\right) + c_p\left(T_{3'} - T_2\right)} \tag{44A}$$

$$= 1 - \left\{ \frac{T_1\left(\frac{T_4}{T_1} - 1\right)}{T_2\left(\frac{T_3}{T_2} - 1\right) + \gamma T_3\left(\frac{T_{3'}}{T_3} - 1\right)} \right\} \tag{44B}$$

$$= 1 - \frac{\frac{T_4}{T_1} - 1}{\frac{T_2}{T_1}\left(\frac{T_3}{T_2} - 1\right) + \frac{\gamma T_3}{T_2}\frac{T_2}{T_1}\left(\frac{T_{3'}}{T_3} - 1\right)} \tag{43C}$$

From thermodynamics

$$\frac{T_3}{T_2} = \frac{p_3}{p_2} = r_p \tag{45}$$

the explosion or pressure ratio and

$$\frac{T_{3'}}{T_3} = \frac{V_{3'}}{V_3} = r_c \tag{46}$$

the cut-off ratio.

Now, $$\frac{T_4}{T_1} = \frac{p_4}{p_1} = \frac{p_4}{p_{3'}} \frac{p_{3'}}{p_3} \frac{p_3}{p_2} \frac{p_2}{p_1}$$

Also $$\frac{p_4}{p_{3'}} = \left(\frac{V_{3'}}{V_4}\right)^\gamma = \left(\frac{V_{3'}}{V_3} \frac{V_3}{V_4}\right)^\gamma = \left(r_c \frac{1}{r}\right)^\gamma$$

And $$\frac{p_2}{p_1} = r^\gamma$$

Thus $$\frac{T_4}{T_1} = r_p r_c^\gamma$$

Also $$\frac{T_2}{T_1} = \left(\frac{V_1}{V_2}\right)^\gamma = r^{\gamma-1}$$

Therefore, the thermal efficiency of the dual cycle is

$$\eta = 1 - \frac{1}{r^{\gamma-1}} \left[\frac{r_p r_c^\gamma - 1}{(r_p - 1) + \gamma r_p (r_c - 1)}\right] \tag{47}$$

We can substitute the value of η from Eq. 46 in Eq. 26 and obtain the value of mep/p_1 for the dual cycle.

In terms of the cut-off ratio and pressure ratio, we can obtain another expression for mep/p_1 as follows:

$$\frac{mep}{p_1} = \frac{\gamma r_p r^\gamma (r_c - 1) + r^\gamma (r_p - 1) - r(r_p r_c^\gamma - 1)}{(r-1)(\gamma - 1)} \tag{48}$$

For the dual cycle, the expression for mep/p_3 is as follows:

$$\frac{mep}{p_3} = \frac{mep}{p_1}\left(\frac{p_1}{p_3}\right) \tag{49}$$

Since the dual cycle is also called the limited pressure cycle, the peak pressure, p_3, is usually specified. Since the initial pressure, p_1, is known, the ratio p_3/p_1 is known. We can correlate r_p with this ratio as follows:

$$r_p = \frac{p_3}{p_1}\left(\frac{1}{r^{\gamma}}\right) \tag{50}$$

We can obtain an expression for r_c in terms of Q' and r_p and other known quantities as follows:

$$r_c = \frac{1}{\gamma}\left(\left[\left\{\frac{Q'}{c_v T_1 r^{\gamma-1}}\right\}\frac{1}{r_p}\right] + (\gamma - 1)\right) \tag{51}$$

We can also obtain an expression for r_p in terms of Q' and r_c and other known quantities as follows:

$$r_p = \frac{\left[\dfrac{Q'}{c_v T_1 r^{\gamma-1}} + 1\right]}{1 + \gamma r_c - \gamma} \tag{52}$$

In a series of air cycles with varying pressure ratio at a given compression ratio and the same Q', the constant volume cycle has the highest efficiency and the constant pressure cycle the lowest efficiency.

The efficiencies of the three cycles for the same value of $Q'\left(\frac{r}{r-1}\right)$ for the same initial conditions and three values of p_3/p_1 for the dual cycle. It is interesting to note that the air standard efficiency is little affected by compression ratio above a compression ratio of 8 for the limited pressure cycle.

The curves of mep/p_3 versus compression ratio for the same three cycles. It is seen that a considerable increase in this ratio is obtained for a limited pressure cycle as compared to the constant volume or constant pressure cycles.

The Atkinson Cycle

This cycle is also referred to as the *complete expansion cycle.* Inspection of the P-V diagrams of the Otto, Diesel and Dual cycles shows that the expansion process to point 4 does not proceed to the lowest possible pressure, namely, atmospheric pressure. This is true of all real engines; when the exhaust valve opens, the high pressure gases undergo a violent blow down process with consequent dissipation of available energy. This is necessary so as to allow the gases to flow out due to pressure difference and hence reduce the piston work in driving out the gases. The air standard cycle shows a loss of net work because of the reduction in area of the P-V diagram.

In the Otto cycle, if the expansion is allowed to completion to point 4' and heat rejection occurs at constant pressure, the cycle is called the Atkinson cycle.

The heat supplied, Q_s per unit mass of charge is given by

$c_v(T_3 - T_2)$

whereas the heat rejected, Q_r per unit mass of charge is given by

$c_p(T_4 - T_1)$

and the thermal efficiency is given by

$$\eta_{th} = 1 - \frac{c_p (T_{4'} - T_1)}{c_v (T_3 - T_2)} \tag{53}$$

$$= 1 - \frac{\gamma T_1 \left(\frac{T_{4'}}{T_1} - 1 \right)}{T_2 \left(\frac{T_3}{T_2} - 1 \right)} \tag{54}$$

Now $$\frac{T_{4'}}{T_1} = \frac{V_{4'}}{V_1} = r_v \tag{54}$$

As before, $$\frac{T_3}{T_2} = \frac{p_3}{p_2} = r_p \tag{55}$$

And $$\frac{T_2}{T_1} = \left(\frac{V_1}{V_2} \right)^{\gamma} = r^{\gamma - 1} \tag{56}$$

The efficiency is therefore given by

$$\eta_{th} = 1 - \frac{1}{r^{\gamma - 1}} \frac{\gamma (r_v - 1)}{(r_p - 1)} \tag{57}$$

If we denote the expansion ratio as $V_{4'}/V_3$, we can rewrite the thermal efficiency as

$$\eta_{th} = 1 - \frac{1}{r_e^{\gamma - 1}} \frac{\gamma (r_v - 1)}{(r_v^{\gamma} - 1)} \tag{58}$$

Since the Atkinson cycle area under the P-V diagram is larger than the corresponding Otto cycle, the efficiency, for the same compression ratio and heat input, will be higher. An engine can be built to make use of complete expansion, but the stroke length of such an engine will be extremely long and will not be economically feasible to offset the improvement in power and efficiency. Also, there are some operational problems with such a cycle.

Actual Working Fluid

So far the analysis has been carried out assuming that the working medium in the engine is air which was assumed to behave as a perfect gas with constant specific heat. In actual fact, the specific heat of air is not constant but varies with temperature. Also, the actual working fluid consists of air and fuel in the proper ratio to achieve the appropriate energy release in order to obtain work from the engine. In addition, the residual gas from the previous cycle will also mix with the fresh charge. The quantity of the residual gas is usually small and may be neglected in certain cases.

This working fluid undergoes compression in the engine. Heat addition will be in the form of combustion of this fluid and result in the formation of products of combustion which has properties that are completely different from the properties of the original air and fuel mixture which would then expand in the engine before being expelled from the cylinder. Thus, the study of a cycle which works on the actual working fluid would first involve studying the thermodynamic properties of the fluid before and after combustion.

Actual Working Fluid before Combustion

Air constitutes a major portion of the actual working fluid in all actual engine cycles. In all reciprocating engines, residual gas is a constituent of the working fluid. Depending on the complexity of the analysis, we can assume the air to be dry. Alternatively, the presence of moisture can be included in the calculations. The presence of fuel depends on whether the engine is a carbureted type (as in the conventional SI engine) or fuel injected type (as in conventional CI engines and fuel injected SI engines).

Air

Dry air contains approximately 23.2% oxygen and 76.8% *atmospheric* nitrogen (which include about 1.2% argon and carbon dioxide). The volumetric (or molar) concentration of oxygen in air is

21% and atmospheric nitrogen 79% (which includes nearly 0.9% argon and carbon dioxide. The molecular weight of air may be taken as 28.96. The ratio of atmospheric nitrogen to oxygen may be taken to be 3.76 by volume and 3.31 by mass.

In addition to the dry elements, air contains water vapour whose concentration varies with temperature and the degree of saturation. The latter varies with time and place from nearly zero to hundred per cent (saturated air). In addition to water vapour, air may contain water droplets in suspension. The quantity of water vapour in a given sample of air may be obtained from psychrometric charts for a given dry bulb temperature and relative humidity or wet bulb temperature. These quantities can be easily measured with the help of suitable instruments. Water vapour increases the characteristic gas constant, R, specific heat at constant volume, c_v and constant pressure, c_p, relative to dry air whereas the adiabatic index, γ, and thermal conductivity, k decrease.

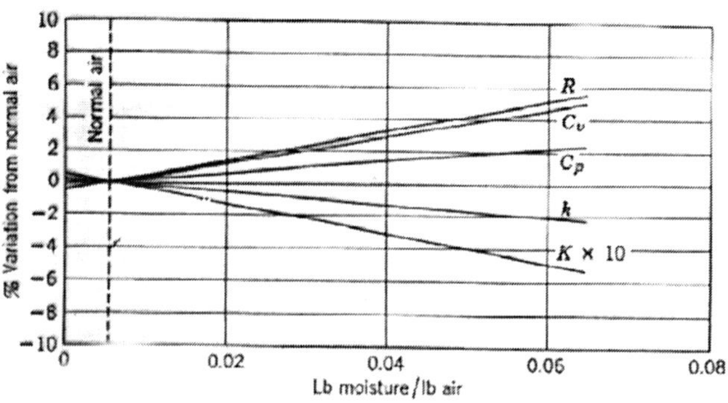

Fig 3-2. Effect of humidity on properties of air: R = universal R/m in eq 2-2; C_v = specific heat at constant volume; C_p = specific heat at constant pressure; k = ratio of specific heats; K = thermal conductivity.

Figure 1: Effect of humidity on properties of air: R=universal R/m in eq 2-2; C_v=specific heat at constant volume; C_p=specific heat at constant pressure; k=ratio of specific heats; K=thermal conductivity.

Thermodynamic Properties of Dry Air

Krieger and Borman proposed the following relation for the internal energy for dry air, after converting to SI units:

$$u = 0.6919943T - 0.3917296\times10^{-4}T^2 + 0.5292534\times10^{-7}T^3 - 0.2286286\times10^{-10}T^4 + 0.277589\times10^{-14}T^5 \quad \text{kJ/kg} \qquad (1)$$

T is the temperature of air in Kelvin.

We can obtain the specific heat at constant volume, c_v, by differentiating the above equation with respect to T, that is

$c_v = du/dT$ (2)

and since R for air is a constant, and equal to 0.287 kJ/kgK, the specific heat at constant pressure, c_p is given by

$c_p = R - c_v$ (3)

A simpler correlation for the specific heat at constant pressure for air is given as follows:

$c_p = 0.9211 + 0.0002306\,T$ kJ/kg-K

T is in Kelvin

A third order equation was proposed by Partha Pratim Saha, 89085,ex student of IITD:

$c_p = 26.430213692 + 8.4435671*10^{-3}T - 2.1567692496*10^{-6}T^2 + 1.9461954*10^{-10}T^3$

The units are kJ $kmole^{-1}$ K^{-1}

T is in Kelvin

Molecular weight of air is 29

According to Lucas, the c_p of any gas is given as follows.

$c_p = a_{ij}\,(T/1000)^{i-1}$

where i = 1to 7 and j represents the particular species, isooctane, oxygen or nitrogen. The units are kJ $kmole^{-1}$ K^{-1}

Table : 1

Species	***Gaseous Isooctane C_8H_{18}***	***Oxygen O_2***	***Nitrogen N_2***
Enthalpy of formation HF at 198.15K(KJ/kgmole)	-224137	0	0
Standard entropy s_0 at 298.15K(KJ/kgmole)	423.21	205.03	191.50
Specificheat Coefficient			
(298-1500K)	a_1	10.6597	30.7990
32.2583	a_2	642.813	-22.4664
-23.0283	a_3	-55.8068	85.9468
54.8998	a_4	-392.337	-108.220
-48.8672	a_5	296.977	66.7167
21.8990	a_6	-69.0190	-20.3296
-4.90861	a_7	0	2.45245
0.434646			

Another correlation is given below.

$$c_p = C_0 + C_1 \frac{T}{1000} + C_2\left(\frac{T}{1000}\right)^2 + C_3\left(\frac{T}{1000}\right)^3 kJ/kgK$$

The values of the coefficients for different gases are given in Table 2 below.

Table: 2

Gas	C_0	C_1	C_2	C_3
Air	1.05	-0.365	0.85	-0.39
Methane	1.2	3.25	0.75	-0.71
CO2	0.45	1.67	-1.27	0.39
Steam	1.79	0.107	0.586	-0.20
O2	0.88	-0.0001	0.54	0.33
N2	1.11	-0.48	0.96	-0.42

Many books give tabulated thermodynamic properties of air as a function of temperature.

Thermodynamic Properties of Fuel

Obert has reproduced enthalpy data for isooctane (2,2,4 trimethyl pentane) in calories per gram above zero Kelvin. These values are tabulated as function of temperature. Empirical equations were obtained by Vivek Saxena, B.Tech. student in 1981, from these data as follows:

$h = A\,(2.0992*10^{-4}*T^{1.838}) - (41200/114.232)$ (For the range of 300-700 K)

$h = A(5.7763*10^{-4}*T^{1.6826}) - (41200/114.232)$ (For the range of 701-1000 K)

where T is in Kelvin. A=1000/114.232. Units of h are kcal/kg

Another formula developed by K.S.Reddy, former research scholar in the department, in 1982, is:

$h = C1T + C2T^{2} + \ldots\ldots\ldots\ldots + C8T^{8} - 172.46$

where the values of C1, C2 are given below.

$C1 = 0.1391491266*10^{-1}$

$C2 = 0.3045267063*10^{-3}$

$C3 = -0.6004287536*10^{-7}$

$C4 = 0.3690244389*10^{-9}$

$C5 = -0.9428759337*10^{-12}$

$C6 = 0.9349048276*10^{-15}$

$C7 = -\ 0.4236288303*10^{-18}$

$C8 = 0.7352647434*10^{-22}$

A formula for specific heat, enthalpy and entropy for some fuels is given by Ferguson. This is given below:

Table : 3

FUEL	a_o	b_o	*Co*	d_o	e_o
CH_3NO_2	1.412633	2.087101E-02	-8.142134E-06	-1.026351E04	1.917126E+01
CH_4	1.971324	7.871586E-03	-1.048592E-06	-9.930422E03	8.873728
CH_3OH	1.779819	1.262503E-02	-3.624890E-06	-2.525420E04	1.50884E+01
C_6H_6	-2.545087	4.79554E-02	-2.030765E-05	8.782234E+03	3.348825E+01
C_7H_{17} [a]	4.0652	6.0977E-02	-1.8801E-05	-3.5880E+04	1.545E+01
$C_{14.4}H_{24.9}$ [b]	7.9710	1.1954E-01	-3.6858E-05	-1.9385E+04	-1.7879

[a] –Gasoline

[b] – Diesel fuel

The constants give the ideal gas properties of the fuel according to

$$C_p/R = a_0 + b_0 T + C_0 T^2$$

$$h/RT = a_0 + (b_0/2)T + (C_0/3)T^2 + d_0/T$$

$$s^0/R = a_0 \ln T + b_0 T + (c_0/2)T^2 + e_0$$

Another Correlation is given below.

$$C_{p,f} = C_0 + C_1 \frac{T}{1000} + C_2\left(\frac{T}{1000}\right)^2 + C_3\left(\frac{T}{1000}\right)^3 + \left(\frac{C_4}{T^2}\right) kJ/$$

The constants for various types of fuels are given in Table 4 below

Table 4.

Fuel	C_0	C_1	C_2	C_3	C_4
Methane	-0.29149	26.327	-10.610	1.5656	0.16573
Propane	-1.4867	74.339	-39.065	8.0543	0.01219
Isooctane	-0.55313	181.62	-97.787	20.402	-0.03095
Gasoline	-24.078	256.63	-201.68	64.750	0.5808
Diesel	-9.1063	246.97	-143.74	32.329	0.0518

Thermodynamic Charts

There are many charts for use in engine analysis. One such set of charts was prepared by Hottel *et al* gives the temperature versus entropy for dry air and air-gasoline (as octane) mixtures and includes the effect of residual gases also.

These charts are in the FPS system of units and values are given over the temperature range of 360-2500R (200-1400K). The mass base of the charts is 1 lb mole of gas. In the chart, the internal energy is assumed to be zero at 560R (100°F) or 311K so that enthalpy is zero at 397R (-63°F) or 220.2K. The values of internal energy, enthalpy and entropy are computed for zero gauge pressure (atmospheric) but the assumption that they hold good over the range encountered with internal combustion engines involves no serious error.

Lichty provided tabulated data of internal energy, enthalpy and entropy as function of temperature for a number of gases including air. The base for these values is zero Rankine. A temperature versus entropy chart provides unburned data for air and three air-fuel mixtures. Lucas prepared a set of charts for isooctane-air mixtures. Five such charts are available, for equivalence ratios 0.8 to 1.2 in steps of 0.1. These charts use thermodynamic data obtained from the JANAF tables. The description of the "Before Combustion" chart is given below.

The "Before Combustion" Charts

The composition of the charge before combustion is assumed to be gaseous isooctane and air. The chemical equation, therefore, is

$$n_1C_8H_{18} + n_2O_2 + n_3N_2 \rightarrow \text{Products} \quad (1)$$

The stoichiometric equation for this reaction is

$$C_8H_{18} + 12.5(O_2 + 3.76N_2) \text{ a } 8CO_2 + 9H_2O + 3.76N_2 \quad (2)$$

The stoichiometric fuel-air ratio of isooctane is 0.066514:1. If the equivalence ratio (also called relative fuel-air ratio or fuel-air equivalence ratio) is denoted by f, the actual fuel-air ratio is 0.066514f, which is the mass of fuel in kg. The mass of the charge is given by 1+0.066514f, that is for 1 kg of air.

The mass equation of the reaction is

$$0.066514f \text{ kg } C_8H_{18} + 1 \text{ kg Air} \rightarrow \text{Products} \quad (3)$$

Now, 1 kg air contains 0.232 kg oxygen of relative atomic mass 32 and 0.768 kg atmospheric nitrogen of relative atomic mass 28.161. The relative molecular mass of isooctane is 114.232. Hence the molar equation is given by

$$0.066514f/114.232\ C_8H_{18}+0.232/32\ O_2+0.768/28.161\ N_2 \rightarrow \text{Products} \quad (4)$$

$$0.000582271f\ C_8H_{18} + 0.00725\ O_2 + 0.0273\ N_2 \rightarrow \text{Products} \quad (5)$$

$$f/1717.42\ C_8H_{18}+12.5/1717.42\ O_2+ (79/21)(12.5/1717.42)\ N_2 \rightarrow \text{Products} \quad (6)$$

Thus, in Eq, 1,n_1 = f/1717.42

n_2 = 12.5/1717.42

n_3 = (79/21)(12.5/1717.42)

Hence, the total number of moles of gaseous reactants is

$$n_{TR} = \sum_{j=1}^{3} n_j$$

$$= \frac{(\varphi + 59.524)}{1717.42}\ moles \qquad (7)$$

The specific heat at constant pressure of a species may be expressed as a sixth order polynomial,

$$c_{pj} = \sum_{i=1}^{7} a_{ij} \left(\frac{T}{1000} \right)^{i-1} kJ / mole \qquad (8)$$

where subscript j denotes the particular species, isooctane, nitrogen or oxygen.

The coefficients, $a_{i,j}$, the enthalpy of formation and the standard entropy data.

The sensible enthalpy, H_s, of the charge, in kJ/kg of air, therefore, is given by

$$H_s = \left\{ \sum_{j=1}^{3} n_j \sum_{i=1}^{7} \frac{a_{ij}}{i} \left[\left(\frac{T}{1000} \right)^i - (0.29815)^i \right] \right\} x1000 \qquad (9)$$

The sensible internal energy, E_s, of the charge in kJ/kg of air, is given by

$$E_s = H_s - n_T\, G(T - 298.15) \qquad (10)$$

where G is the universal gas constant. The entropy, S, of the charge in kJ/K per kg of air, is given by

$$S = \sum_{j=1}^{3} n_j \left[a_1 \ln \frac{T}{298.15} + \sum_{i=2}^{7} \left(\frac{a_{ij} \left(\left[\frac{T}{1000} \right]^{i-1} - [0.29815]^{i-1} \right)}{i-1} \right) + S_{oj} - G \ln \left(\right. \right.$$

(11)

Here, p_j is the partial pressure of the species and p_o is the datum pressure of 101.35 kN/m^2 (1 atmosphere).

The residual charge is taken into account during combustion. The enthalpy of the residual gas is given by

$$h_{res\,gas} = fn_{TP}\sum_{j=1}^{m_p}\left(x_j HF_j\right) \tag{12}$$

where n_{TP} is the total number of moles of products,

f is the residual gas fraction,

m_p is the number of the product species in the residual gas,

x_j is the concentration of each of these species, and

HF_j is the corresponding enthalpy.

The Lucas chart does not take into account the presence of moisture in the air. In some cases, the presence of moisture is included in the residual gas fraction. It is assumed, based on certain observations, that with the fractions of water vapour normally associated with the atmosphere, the internal energy and enthalpy are the same as those of the products of combustion of a fuel-air mixture with the following characteristics:

$$F_R = 3.54h \tag{13}$$

where h is the humidity ratio, or the mass of water vapour associated with unit mass of dry air and F_R is the fuel-air equivalence ratio, φ. The actual equivalence ratio, taking into account the moisture $F_{v,R}$, is given by

$$F_{v,R} = F_R + 3.54h \tag{14}$$

$$f_v = f + \frac{3.54h}{F_R + 3.54h} \tag{15}$$

and $F_{v,R}$ and f_v are the values to be used in reading the values in the charts where applicable. F_R and f are actual values based on dry air.

Relation between Unburned and Burned Mixture Charts

Given the thermodynamic properties of the unburned mixture before combustion, we would like to know the state of the burned mixture following adiabatic (i) constant-volume and (ii) constant pressure combustion. The datum for the internal energy and enthalpy for the unburned mixture is zero internal energy and enthalpy at 298.15 K. For the burned mixture, zero enthalpy for the gaseous species oxygen and nitrogen at 298.15 K is assumed.

If $\Delta h^{o}_{f,u}$ is the enthalpy of formation of the unburned mixture at 298.15 K, per kilogram of air in the original mixture and $\Delta\%^{o}_{f,i}$ is the enthalpy of formation of species i at 298.15 K, per kilomole, then

$$\Delta h^{o}_{f,u} = \Sigma n_i \Delta\%^{o}_{f,i}$$

where n_i is the number of kilomoles of species i per kilogram of air. The unburned mixture enthalpy, h_u, with the same datum as the unburned mixture enthalpy, is therefore given by the sensible enthalpy, $h_{s,u}$ and $\Delta h^o_{f,u}$, thus

$$h_u = h_{s,u} + \Delta h^o_{f,u}$$

and similarly, the internal energy u_u is given by

$$u_u = u_{s,u} + \Delta u^o_{f,u}$$

and $\Delta u^o_{f,u}$ can be obtained either from the equation

$$\Delta u^o_{f,u} = \Sigma n_i \Delta \hat{u}^o_{f,i}$$

or from $\Delta u^o_{f,u}$ thus

$$\Delta u^o_{f,u} = \Delta h^o_{f,u} - (n_P - n_R)R_o T$$

For constant volume adiabatic combustion, $u_u = u_b$ and for constant pressure adiabatic combustion, $h_u = h_b$.

The section below explains how the unburned and burned enthalpy at the datum can be obtained and from that the enthalpy at the appropriate temperature can be obtained. This procedure gives the link between the unburned (or "before combustion") and the burned (or "after combustion") mixture chart.

Chemical Internal Energy

The incoming charge is deemed to consist of air plus residual combustion products left over from the previous cycle. If the volumetric residual gas is f, the composition of the new charge may be written as

$$\frac{(1-f)}{1717.42}\{\phi.C_8H_{18} + 12.5\,O_2 + 47.024\,N_2\} + f.n_{TP}\left\{\sum_{j=1}^{m_p} x_j A_p\right.$$

where n_{TP} is the total number of moles of the product gases resulting from the combustion of fuel in 1 kg of air, m_p is the number of product species $A_{p,j}$ and x_j is the concentration of each of these species.

This means that the nominal 1 kg of air is contaminated with a small proportion of residual gas.

At the datum temperature of 298.15 K, therefore, the enthalpy of the charge per kg of air is

$$h_c = \frac{(1-f)}{1717.42}\phi(-224137) + f\,n_{TP}\left\{\sum_{j=1}^{m_p}\left(x_j HF_j\right)\right\}$$

The enthalpy of formation of isooctane has been taken from the table and the concentrations of the main product species together with the sum of their enthalpy values are also given in the table.

At any other temperature, the enthalpy of the charge is given by

$$h = h_c + \sum_{k=1}^{m_R} n_k \bar{c}_{p_k} (T - 298.15)$$

where m_R is the number of species in the charge, that is, $h = h_c + h_s$ where h_s is the sensible enthalpy as shown above.

The internal energy of the charge per kg of air at the datum temperature of 298.15 K is

$$e_c = h_c - 298.15xG\{(1 - f)n_{TR} + fn_{TP}\}$$

where the term in the brackets is the total number of moles of the charge per nominal 1 kg of air and G is the universal gas constant.

At any other temperature T, the internal energy is given by $e = e_c + e_s$ so that

$$e = e_c + \sum_{k=1}^{m_R} n_k \bar{c}_{v_k} (T - 298.15)$$

The constants for the evaluation for the chemical internal energy, e_c. The sensible internal energy, e_s, may be read from the vertical axis of the "before combustion" charts.

The "After Combustion" Charts

The after combustion charts were obtained, taking dissociation into account, assuming that thirteen species to exist in the products. These are H_2O, H_2, H, O_2, O, OH, CO_2, CO, N_2, N, NO, NO_2, and N_2O. The last two oxides are taken because exhaust pollution research has shown them to exist, albeit in small quantities. Their absence makes little difference to the charts.

The after combustion charts are plots of internal energy against entropy for lines of constant pressure, constant temperature and constant volume. In the low temperature range, where the constant temperature lines are practically horizontal, a subsidiary scale of sensible enthalpy, h_s, may be constructed on the vertical axis to some advantage. This will help in estimating the temperature of the new charge after it has mixed with the residual.

The dissociation calculations at a specified temperature and pressure were performed by using the Newton-Raphson iteration technique to solve simultaneously the mass balance and equilibrium constant equations.

The temperature range of the after combustion charts is 600 to 4000 K, the pressure range is from 20 to 100,000 kPa and the volume range from 0.01 to 10.0 m^3.

Thermodynamic Relations for Engine Processes

The overall engine operating parameters of greatest interest which can be determined from a thermodynamic analysis of the engine are:

The indicated fuel conversion efficiency $\eta_{f,i}$:

$$\eta_{f,i} = \frac{W_{c,i}}{m_f Q_{cv}}$$

which, since the combustion efficiency is assumed to be unity, is equal to the indicated thermal efficiency (or indicated thermal conversion efficiency), $\eta_{th,i}$, and the indicated mean effective pressure (imep) given by

$$\text{imep} = \frac{W_{c,i}}{V_d} = \frac{\eta_{f,i} m_f Q_{cv}}{V_d}$$

where $W_{c,i}$, the indicated work in kg of air per cycle, is the sum of the compression stroke work and the expansion stroke work:

$$W_{c,i} = W_C + W_E$$

We can use the first and second laws of thermodynamics to the cylinder contents to obtain the following relationships:

Compression stroke:

$$r_c = \frac{v_1}{v_2}$$

Since the process is reversible and adiabatic,

$$s_1 = s_2$$

The compression work, per kg of air is

$$W_C = u_1 - u_2$$

Combustion process:

For the constant volume cycle,

$$v_3 = v_2 \quad u_3 = u_2$$

For the constant pressure cycle,

$$p_3 = p_2 \quad h_3 = h_2$$

For the limited pressure cycle,

$$v_3 = v_2 \quad u_3 = u_2$$

$$p_{3a} = p_3 \quad h_{3a} = h_3$$

The heat added per kg of air is given by FQ_{cv} where F is the fuel-air ratio, in kg fuel per kg air used in one cycle. Q_{cv} is the higher heating value of the fuel per kg of fuel per cycle.

Expansion stroke:

For the constant volume cycle,

$$r_c = \frac{v_4}{v_3}$$

Since the process is reversible and adiabatic,

$$s_4 = s_3$$

The expansion work, per kg of air is

$$W_E = u_3 - u_4$$

For the constant pressure cycle,

$$p_3 = p_2 \quad r_c = \frac{v_4}{v_2} \quad s_4 = s_3$$

and the expansion work per kg of air is,

$$W_E = u_3 - u_4 + p_2(v_3 - v_2)$$

$$= h_3 - h_4 + p_4v_4 - p_2v_2$$

For the limited pressure cycle,

$$r_c = \frac{v_4}{v_{3a}} \quad p_{3a} = p_3 \quad s_4 = s_{3a}$$

and the expansion work is

$$W_E = u_{3a} - u_4 + p_3(v_{3a} - v_3)$$

$$= h_{3a} - h_4 + p_4v_4 - p_3v_3$$

The indicated fuel conversion efficiency of the three cycles is found by substituting for indicated work per cycle and the heat added by the fuel:

For the constant volume cycle:

$$\eta_{f,i} = \frac{(u_3 - u_4) - (u_2 - u_1)}{FQ_{cv}}$$

$$= \frac{(u_1 - u_2 + u_3 - u_4)}{FQ_{cv}}$$

For the constant pressure cycle:

$$\eta_{f,i} = \frac{(h_3 - h_4) - (u_2 - u_1) + p_4 v_4 - p_2 v_2}{FQ_{cv}}$$

For the limited pressure cycle:

$$\eta_{f,i} = \frac{(h_{3a} - h_4) - (u_2 - u_1) + p_4 v_4 - p_3 v_3}{FQ_{cv}}$$

As before, the state of the mixture at point 1 in the cycle depends on the intake mixture properties and the residual gas properties at the end of the exhaust stroke.

Since $W_{c,i}$ is the indicated work per kg of air per cycle, its reciprocal is the indicated specific air consumption, *isac* in appropriate units, thus

$$isac = \frac{1}{W_{c,i}}$$

and the indicated specific fuel consumption, *isfc* in appropriate units, is given by

$$\text{isfc} = \text{F.isac}$$

Lubrication

Figure 2: Lubrication of the ship steam engine crankshaft. The two bottles with lubricant are attached to the piston and move while the engine is operating.

Lubrication is the process, or technique employed to reduce wear of one or both surfaces in close proximity, and moving relative to each another, by interposing a substance called lubricant between the surfaces to carry or to help carry the load (pressure generated) between the opposing surfaces. The interposed lubricant film can be a solid, (e.g. graphite, MoS_2) a solid/liquid dispersion, a liquid, a liquid-liquid dispersion gasses exceptionally a gas.

In the most common case the applied load is carried by pressure generated within the fluid due to the frictional viscous resistance to motion of the lubricating fluid between the surfaces.

Lubrication can also describe the phenomenon such reduction of wear occurs without human intervention (hydroplaning on a road).

The science of friction, lubrication and wear is called tribology.

Adequate lubrication allows smooth continuous operation of equipment, with only mild wear, and without excessive stresses or seizures at bearings. When lubrication breaks down, metal or other components can rub destructively over each other, causing destructive damage, heat, and failure.

The Regimes of Lubrication

As the load increases on the contacting surfaces three distinct situations can be observed with respect to the mode of lubrication, which are called regimes of lubrication:

- Fluid film lubrication is the lubrication regime in which through viscous forces the load is fully supported by the lubricant within the space or gap between the parts in motion relative to one another (the lubricated conjunction) and solid–solid contact is avoided.
    - Hydrostatic lubrication is when an external pressure is applied to the lubricant in the bearing, to maintain the fluid lubricant film where it would otherwise be squeezed out.
    - Hydrodynamic lubrication is where the motion of the contacting surfaces, and the exact design of the bearing is used to pump lubricant around the bearing to maintain the lubricating film. This design of bearing may wear when started or stopped, as the lubricant film breaks down.
- Elastohydrodynamic lubrication: The opposing surfaces are separated, but there occurs some interaction between the raised

solid features called *asperities*, and there is an elastic deformation on the contacting surface enlarging the load-bearing area whereby the viscous resistance of the lubricant becomes capable of supporting the load.

- Boundary lubrication (also called boundary film lubrication): The bodies come into closer contact at their asperities; the heat developed by the local pressures causes a condition which is called stick-slip and some asperities break off. At the elevated temperature and pressure conditions chemically reactive constituents of the lubricant react with the contact surface forming a highly resistant tenacious layer, or film on the moving solid surfaces (boundary film) which is capable of supporting the load and major wear or breakdown is avoided. Boundary lubrication is also defined as that regime in which the load is carried by the surface asperities rather than by the lubricant.

Besides supporting the load the lubricant may have to perform other functions as well, for instance it may cool the contact areas and remove wear products. While carrying out these functions the lubricant is constantly replaced from the contact areas either by the relative movement (hydrodynamics) or by externally induced forces.

Lubrication is required for correct operation of mechanical systems pistons, pumps, cams, bearings, turbines, cutting tools etc. where without lubrication the pressure between the surfaces in close proximity would generate enough heat for rapid surface damage which in a coarsened condition may literally weld the surfaces together, causing seizure.

In some applications, such as piston engines, the film between the piston and the cylinder wall also seals the combustion chamber, preventing combustion gases from escaping into the crankcase.

Skin Friction

Skin friction arises from the friction of the fluid against the "skin" of the object that is moving through it. Skin friction arises from the interaction between the fluid and the skin of the body, and is directly related to the area of the surface of the body that is in contact with the fluid. Skin friction follows the drag equation and rises with the square of the velocity.

Skin friction is caused by viscous drag in the boundary layer around the object. There are two ways to decrease skin friction: the first is to shape the moving body so that smooth flow is possible, like an airfoil. The second method is to decrease the length and cross-section of the moving object as much as is practicable.

Internal Friction

Internal friction is the force resisting motion between the elements making up a solid material while it undergoes plastic deformation.

Plastic deformation in solids is an irreversible change in the internal molecular structure of an object. This change may be due to either (or both) an applied force or a change in temperature. The change of an object's shape is called strain. The force causing it is called stress. Stress does not necessarily cause permanent change. As deformation occurs, internal forces oppose the applied force. If the applied stress is not too large these opposing forces may completely resist the applied force, allowing the object to assume a new equilibrium state and to return to its original shape when the force is removed. This is what is known in the literature as elastic deformation (or elasticity). Larger forces in excess of the elastic limit may cause a permanent (irreversible) deformation of the object. This is what is known as plastic deformation.

Other Types of Friction

Rolling Resistance

Rolling resistance is the force that resists the rolling of a wheel or other circular object along a surface caused by deformations in the object and/or surface. Generally the force of rolling resistance is less than that associated with kinetic friction. Typical values for the coefficient of rolling resistance are 0.001. One of the most common examples of rolling resistance is the movement of motor vehicle tires on a road, a process which generates heat and sound as by-products.

Triboelectric Effect

Rubbing dissimilar materials against one another can cause a build-up of electrostatic charge, which can be hazardous if flammable gases or vapours are present. When the static build-up discharges, explosions can be caused by ignition of the flammable mixture.

Belt Friction

Belt friction is a physical property observed from the forces acting on a belt wrapped around a pulley, when one end is being pulled. The resulting tension, which acts on both ends of the belt, can be modelled by the belt friction equation.

In practice, the theoretical tension acting on the belt or rope calculated by the belt friction equation can be compared to the maximum tension the belt can support. This helps a designer of such a rig to know how many times the belt or rope must be wrapped around the pulley to prevent it from slipping. Mountain climbers and sailing crews demonstrate a standard knowledge of belt friction when accomplishing basic tasks.

Reducing Friction

Devices

Devices such as wheels, ball bearings, roller bearings, and air cushion or other types of fluid bearings can change sliding friction into a much smaller type of rolling friction.

Many thermoplastic materials such as nylon, HDPE and PTFE are commonly used in low friction bearings. They are especially useful because the coefficient of friction falls with increasing imposed load. For improved wear resistance, very high molecular weight grades are usually specified for heavy duty or critical bearings.

Lubricants

A common way to reduce friction is by using a lubricant, such as oil, water, or grease, which is placed between the two surfaces, often dramatically lessening the coefficient of friction. The science of friction and lubrication is called tribology. Lubricant technology is when lubricants are mixed with the application of science, especially to industrial or commercial objectives.

Superlubricity, a recently-discovered effect, has been observed in graphite: it is the substantial decrease of friction between two sliding objects, approaching zero levels. A very small amount of frictional energy would still be dissipated.

Lubricants to overcome friction need not always be thin, turbulent fluids or powdery solids such as graphite and talc; acoustic lubrication actually uses sound as a lubricant.

Another way to reduce friction between two parts is to superimpose micro-scale vibration to one of the parts. This can be sinusoidal vibration as used in ultrasound-assisted cutting or vibration noise, known as dither.

Energy of friction

According to the law of conservation of energy, no energy is destroyed due to friction, though it may be lost to the system of concern. Energy is transformed from other forms into heat. A sliding hockey puck comes to rest because friction converts its kinetic energy into heat. Since heat quickly dissipates, many early philosophers, including Aristotle, wrongly concluded that moving objects lose energy without a driving force.

When an object is pushed along a surface, the energy converted to heat is given by:

$$E_{th} = \mu_k \int F_n(x)dx$$

where

F_n is the normal force,

μ_k is the coefficient of kinetic friction,

x is the coordinate along which the object transverses.

Energy lost to a system as a result of friction is a classic example of thermodynamic irreversibility.

Work of Friction

In the reference frame of the interface between two surfaces, static friction does *no* work, because there is never displacement between the surfaces. In the same reference frame, kinetic friction is always in the direction opposite the motion, and does *negative* work. However, friction can do *positive* work in certain frames of reference. One can see this by placing a heavy box on a rug, then pulling on the rug quickly. In this case, the box slides backwards relative to the rug, but moves forward relative to the frame of reference in which the floor is stationary. Thus, the kinetic friction between the box and rug accelerates the box in the same direction that the box moves, doing *positive* work.

The work done by friction can translate into deformation, wear, and heat that can affect the contact surface properties (even the coefficient of friction between the surfaces). This can be beneficial as in polishing. The work of friction is used to mix and join materials such as in the process of friction welding. Excessive erosion or wear of mating surfaces occur when work due frictional forces rise to unacceptable levels. Harder corrosion particles caught between mating surfaces (fretting) exacerbates wear of frictional forces. Bearing seizure or failure may result from excessive wear due to work of friction. As surfaces are worn by work due to friction, fit and surface finish of an object may degrade until it no longer functions properly.

Applications

Friction is an important factor in many engineering disciplines.

Transportation

- Rail adhesion refers to the grip wheels of a train have on the rails.
- Road slipperiness is an important design and safety factor for automobiles
  - o Split friction is a particularly dangerous condition arising due to varying friction on either side of a car.
  - o Road texture affects the interaction of tires and the driving surface.

Measurement

- A tribometre is an instrument that measures friction on a surface.
- A profilograph is a device used to measure pavement surface roughness.

Wear

In materials science, wear is erosion or sideways displacement of material from its "derivative" and original position on a solid surface performed by the action of another surface.

Wear is related to interactions between surfaces and more specifically the removal and deformation of material on a surface as a result of mechanical action of the opposite surface. The need for relative motion between two surfaces and initial mechanical contact between asperities is an important distinction between mechanical wear compared to other processes with similar outcomes.

The definition of wear may include loss of dimension from plastic deformation if it is originated at the interface between two sliding surfaces.

However, plastic deformation such as yield stress is excluded from the wear definition if it doesn't incorporates a relative sliding motion and contact against another surface despite the possibility for material removal, because it then lacks the relative sliding action of another surface.

Impact wear is in reality a short sliding motion where two solid bodies interact at an exceptional short time interval. Previously due to the fast execution, the contact found in impact wear was referred to as an impulse contact by the nomenclature. Impulse can be described as a mathematical model of a synthesised average on the energy transport between two travelling solids in opposite converging contact.

Cavitation wear is a form of wear where the erosive medium or counter-body is a fluid. Corrosion may be included in wear phenomenons, but the damage is amplified and performed by chemical reactions rather than mechanical action.

Wear can also be defined as a process where interaction between two surfaces or bounding faces of solids within the working environment results in dimensional loss of one solid, with or without any actual decoupling and loss of material. Aspects of the working environment which affect wear include loads and features such as unidirectional sliding, reciprocating, rolling, and impact loads, speed, temperature, but also different types of counter-bodies such as solid, liquid or gas and type of contact ranging between single phase or multiphase, in which the last multiphase may combine liquid with solid particles and gas bubbles.

Measurement

Today there exist a couple of standard test methods for different types of wear to determine the amount of material removal during a specified time period under well-defined conditions.

The ASTM International Committee G-2 attempts to standardise wear testing for specific applications, which are periodically updated. The Society for Tribology and Lubrication Engineers (STLE) has documented a large number of frictional wear and lubrication tests. But all test methods have inbuilt limitations and do not give a true picture in every aspect.

This can be attributed to the complex nature of wear, in particular "industrial wear", and the difficulties associated with accurately simulating wear processes.

A standard result review for wear tests, defined by the ASTM International and respective subcommittees such as Committee G-2, should be expressed as loss of material during wear in terms of volume. The volume loss gives a truer picture than weight loss, particularly when comparing the wear resistance properties of materials with large differences in density.

For example, a weight loss of 14 g in a sample of tungsten carbide + cobalt (density = 14000 kg/m^3) and a weight loss of 2.7 g in a similar sample of aluminium alloy (density = 2700 kg/m^3) both result in the same level of wear (1 cm^3) when expressed as a volume loss. The inverse of volume loss can be used as a comparable index of wear resistance.

Standard wear tests are only used for comparative material ranking of a specific test parameter as stipulated in the test method. For more realistic values of material deterioration in industrial applications it is necessary to conduct wear testing under conditions simulating the exact wear process.

The working life of an engineering component is expired when dimensional losses exceed the specified tolerance limits. Wear, along with other ageing processes such as fatigue and creep in association with stress concentration factors such as fracture toughness causes materials to progressively degrade, eventually leading to material failure at an advanced age.

Wear in industrial applications is one of a limited number of fault factors in which an object loses its usefulness and the economic implication can be of enormous value to the industry.

Stages of Wear

Under normal mechanical and practical procedures, the wear-rate normally changes through three different stages:

- Primary stage or early run-in period, where surfaces adapt to each other and the wear-rate might vary between high and low.
- Secondary stage or mid-age process, where a steady rate of ageing is in motion. Most of the components operational life is comprised in this stage.
- Tertiary stage or old-age period, where the components are subjected to rapid failure due to a high rate of ageing.

The secondary stage is shortened with increasing severity of environmental conditions such as higher temperatures, strain rates, stress and sliding velocities etc.

In explicit wear tests simulating industrial conditions between metallic surfaces, there are no clear chronological distinction between different wear-stages due to big overlaps and symbiotic relations between various friction mechanisms. Surface engineering and treatments are used to minimise wear and extend the components working life.

Types

The study of the processes of wear is part of the discipline of tribology. The complex nature of wear has delayed its investigations and resulted in isolated studies towards specific wear mechanisms or processes. Some commonly referred to wear mechanisms (or processes) include:

1. Adhesive wear
2. Abrasive wear
3. Surface fatigue
4. Fretting wear
5. Erosive wear.

A number of different wear phenomena are also commonly encountered and presented in the literature. Impact-, cavitation-, diffusive- and corrosive- wear are all such examples.

These wear mechanisms, however, do not necessarily act independently and wear mechanisms are not mutually exclusive. "Industrial Wear" are commonly described as incidence of multiple wear mechanisms occurring in unison. Another way to describe "Industrial Wear" is to define clear distinctions in how different friction mechanisms operate, for example distinguish between mechanisms with high or low energy density. Wear mechanisms and/or sub-mechanisms frequently overlap and occur in a synergistic manner, producing a greater rate of wear than the sum of the individual wear mechanisms.

Adhesive Wear

Adhesive wear can be found between surfaces during frictional contact and generally refers to unwanted displacement and attachment of wear debris and material compounds from one surface to another. Two separate mechanisms operate between the surfaces.

1. Adhesive wear are caused by relative motion, "direct contact" and plastic deformation which create wear debris and material transfer from one surface to another.
2. Cohesive adhesive forces, holds two surfaces together even though they are separated by a measurable distance, with or without any actual transfer of material.

The above description and distinction between "Adhesive wear" and its Counterpart "cohesive adhesive forces" are quite common.

Usually cohesive surface forces and adhesive energy potentials between surfaces are examined as a special field in physic departments.

The adhesive wear and material transfer due to direct contact and plastic deformation are examined in engineering science and in industrial research.

Two aligned surfaces may always cause material transfer and due to overlaps and symbiotic relations between relative motional "wear" and "chemical" cohesive attraction, the wear-categorisation have been a source for discussion. Consequently, the definitions and nomenclature must evolve with the latest science and empiric observations.

Generally, adhesive wear occurs when two bodies slide over or are pressed into each other, which promote material transfer. This can be described as plastic deformation of very small fragments within the surface layers. The asperities or microscopic high points or surface roughness found on each surface, define the severity on how fragments of oxides are pulled off and adds to the other surface, partly due to strong adhesive forces between atoms but also due to accumulation of energy in the plastic zone between the asperities during relative motion.

The outcome can be a growing roughening and creation of protrusions (i.e., lumps) above the original surface, in industrial manufacturing referred to as galling , which eventually breaches the oxidised surface layer and connects to the underlying bulk material which enhance the possibility for a stronger adhesion and plastic flow around the lump.

The geometry and the nominal sliding velocity of the lump defines how the flowing material will be transported and accelerated around the lump which is critical to define contact pressure and developed temperature during sliding. The mathematical function for acceleration of flowing material is thereby defined by the lumps surface contour.

It's clear, given these prerequisites, that contact pressure and developed temperature is highly dependent on the lumps geometry.

Flow of material exhibits an increase in energy density, because initial phase transformation and displacement of material demand acceleration of material and high pressure.

Low pressure is not compatible with plastic flow, only after deceleration may the flowing material be exposed to low pressure and quickly cooled. In other words, you can't deform a solid material using direct contact without applying a high pressure and somewhere along the process must acceleration and deceleration take place, i.e., high pressure must be applied on all sides of the deformed material. Flowing material will immediately exhibit energy loss and reduced ability to flow due to phase transformation, if ejected from high pressure into low pressure.

This ability withholds the high pressure and energy density in the contact zone and decreases the amount of energy or friction force needed for further advancement when the sliding continues and partly explain the difference between the static and sliding coefficient of friction (μ) if the main fracture mechanisms are equal to the previous.

Adhesive wear is a common fault factor in industrial applications such as sheet metal forming (SMF) and commonly encountered in conjunction with lubricant failures and are often referred to as welding wear or galling due to the exhibited surface characteristics, phase transition and plastic flow followed by cooling.

The type of mechanism and the amplitude of surface attraction, varies between different materials but are amplified by an increase in the density of "surface energy". Most solids will adhere on contact to some extent. However, oxidation films, lubricants and contaminants naturally occurring generally suppress adhesion and spontaneous exothermic chemical reactions between surfaces generally produce a substance with low energy status in the absorbed species.

Abrasive Wear

Abrasive wear occurs when a hard rough surface slides across a softer surface. ASTM (American Society for Testing and Materials) defines it as the loss of material due to hard particles or hard protuberances that are forced against and move along a solid surface.

Abrasive wear is commonly classified according to the type of contact and the contact environment. The type of contact determines the mode of abrasive wear. The two modes of abrasive wear are known as two-body and three-body abrasive wear. Two-body wear occurs when the grits or hard particles remove material from the opposite surface.

The common analogy is that of material being removed or displaced by a cutting or plowing operation. Three-body wear occurs when the particles are not constrained, and are free to roll and slide down a surface. The contact environment determines whether the wear is classified as open or closed. An open contact environment occurs when the surfaces are sufficiently displaced to be independent of one another.

There are a number of factors which influence abrasive wear and hence the manner of material removal. Several different mechanisms have been proposed to describe the manner in which the material is removed. Three commonly identified mechanisms of abrasive wear are:

1. Plowing
2. Cutting
3. Fragmentation.

Plowing occurs when material is displaced to the side, away from the wear particles, resulting in the formation of grooves that do not involve direct material removal. The displaced material forms ridges adjacent to grooves, which may be removed by subsequent passage of abrasive particles. Cutting occurs when material is separated from the surface in the form of primary debris, or microchips, with little or no material displaced to the sides of the grooves. This mechanism closely resembles conventional machining. Fragmentation occurs when material is separated from a surface by a cutting process and the indenting abrasive causes localised fracture of the wear material. These cracks then freely propagate locally around the wear groove, resulting in additional material removal by spalling.

Abrasive wear can be measured as loss of mass by the Taber Abrasion Test according to ISO 9352 or ASTM D 1044.

Surface Fatigue

Surface fatigue is a process by which the surface of a material is weakened by cyclic loading, which is one type of general material fatigue. Fatigue wear is produced when the wear particles are detached by cyclic crack growth of microcracks on the surface. These microcracks are either superficial cracks or subsurface cracks.

Fretting Wear

Fretting wear is the repeated cyclical rubbing between two surfaces, which is known as fretting, over a period of time which will remove material from one or both surfaces in contact. It occurs typically in bearings, although most bearings have their surfaces hardened to resist the problem. Another problem occurs when cracks in either surface are created, known as fretting fatigue. It is the more serious of the two phenomena because it can lead to catastrophic failure of the bearing. An associated problem occurs when the small particles removed by wear are oxidised in air. The oxides are usually harder than the underlying metal, so wear accelerates as the harder particles abrade the metal surfaces further. Fretting corrosion acts in the same way, especially when water is present. Unprotected bearings on large structures like bridges can suffer serious degradation in behaviour, especially when salt is used during winter to deice the highways carried by the bridges. The problem of fretting corrosion was involved in the Silver Bridge tragedy and the Mianus River Bridge accident.

Erosive Wear

Erosive wear can be described as an extremely short sliding motion and is executed within a short time interval. Erosive wear is caused by the impact of particles of solid or liquid against the surface of an object. The impacting particles gradually remove material from the surface through repeated deformations and cutting actions. It is a widely encountered mechanism in industry. A common example is the erosive wear associated with the movement of slurries through piping and pumping equipment.

The rate of erosive wear is dependent upon a number of factors. The material characteristics of the particles, such as their shape, hardness, impact velocity and impingement angle are primary factors along with the properties of the surface being eroded. The impingement angle is one of the most important factors and is widely recognised in literature. For ductile materials the maximum wear rate is found when the impingement angle is approximately 30°, whilst for non ductile materials the maximum wear rate occurs when the impingement angle is normal to the surface.

Heat Transfer

Heat transfer is a discipline of thermal engineering that concerns the exchange of thermal energy from one physical system to another. Heat transfer is classified into various mechanisms, such as heat conduction, convection, thermal radiation, and phase-change transfer. All forms of heat transfer may occur in some systems (for example, in transparent fluids like the Earth's atmosphere) at the same time. Heat transfer only occurs because of a temperature-difference driving force and heat flows from the high to the low temperature region.

Heat conduction, also called diffusion, is the direct microscopic exchange of kinetic energy of particles through the stationary boundary between two systems. Heat conduction occurs between stationary masses where there is no movement to carry heat away. Heat transfer through the stationary air layer immediately adjacent to, say one millimetre or so from an interior wall, or from a warm pot placed on a counter, are examples of conductive heat transfer. Heat convection occurs when bulk flow of a fluid (gas or liquid) carries heat along with the flow of matter in the fluid. The flow of fluid may be forced by external processes, or sometimes (in gravitational fields) by buoyancy forces caused when thermal energy expands the fluid (for example in a fire plume), thus influencing its own transfer. The latter process is sometimes called "natural convection". All convective processes also move heat partly by diffusion, as well. Another form of convection is forced convection. In this case the fluid is forced to flow by use of a pump, fan or other mechanical means.

The final major form of heat transfer is by radiation, which occurs in any transparent medium (solid or fluid) but may also even occur across vacuum (as when the Sun heats the Earth). Radiation is the transfer of energy through space by means of electromagnetic waves in much the same way as electromagnetic light waves transfer light. The same laws that govern the transfer of light govern the radiant transfer of heat.

Overview

Heat is defined in physics as the transfer of thermal energy across a well-defined boundary around a thermodynamic system. It is a characteristic of a process and is not statically contained in matter. In engineering contexts, however, the term *heat transfer* has acquired a specific usage, despite its literal redundancy of the characterisation of transfer. In these contexts, *heat* is taken as synonymous to thermal energy. This usage has its origin in the historical interpretation of heat as a fluid (*caloric*) that can be transferred by various causes, and that is also common in the language of laymen and everyday life.

Fundamental methods of heat transfer in engineering include conduction, convection, and radiation. Physical laws describe the behaviour and characteristics of each of these methods. Real systems often exhibit a complicated combination of them. Heat transfer methods are used in numerous disciplines, such as automotive engineering, thermal management of electronic devices and systems, climate control, insulation, materials processing, and power plant engineering.

Various mathematical methods have been developed to solve or approximate the results of heat transfer in systems. Heat transfer is a path function (or process quantity), as opposed to a state quantity; therefore, the amount of heat transferred in a thermodynamic process that changes the state of a system depends on how that process occurs, not only the net difference between the initial and final states of the process. Heat flux is a quantitative, vectorial representation of the heat flow through a surface.

Heat transfer is typically studied as part of a general chemical engineering or mechanical engineering curriculum. Typically, thermodynamics is a prerequisite for heat transfer courses, as the laws of thermodynamics are essential to the mechanism of heat transfer. Other courses related to heat transfer include energy conversion, thermofluids, and mass transfer. The transport equations for thermal energy (Fourier's law), mechanical momentum (Newton's law for fluids), and mass transfer (Fick's laws of diffusion) are similar and analogies among these three transport processes have been developed to facilitate prediction of conversion from any one to the others.

Conduction

On a microscopic scale, heat conduction occurs as hot, rapidly moving or vibrating atoms and molecules interact with neighbouring atoms and molecules, transferring some of their energy (heat) to these neighbouring particles. In other words, heat is transferred by conduction when adjacent atoms vibrate against one another, or as electrons move from one atom to another. Conduction is the most significant means of heat transfer within a solid or between solid objects in thermal contact. Fluids—especially gases—are less conductive. Thermal contact conductance is the study of heat conduction between solid bodies in contact.

Steady state conduction is a form of conduction that happens when the temperature difference driving the conduction is constant, so that after an equilibration time, the spatial distribution of temperatures in the conducting object does not change any further. In steady state conduction, the amount of heat entering a section is equal to amount of heat coming out. *Transient conduction* occurs when the temperature within an object changes as a function of time. Analysis of transient systems is more complex and often calls for the application of approximation theories or numerical analysis by computer.

Convection

Convective heat transfer, or convection, is the transfer of heat from one place to another by the movement of fluids. (In physics, the term *fluid* means any substance that deforms under shear stress; it includes liquids, gases, plasmas, and some plastic solids.) Bulk motion of the fluid enhances the heat transfer between the solid surface and the fluid. Convection is usually the dominant form of heat transfer in liquids and gases. Although often discussed as a third method of heat transfer, convection actually describes the combined effects of conduction and fluid flow. Free, or natural, convection occurs when the fluid motion is caused by buoyancy forces that result from density variations due to variations of temperature in the fluid. *Forced* convection is when the fluid is forced to flow over the surface by external means—such as fans, stirrers, and pumps—creating an artificially induced convection current.

Convective heating or cooling in some circumstances may be described by Newton's law of cooling: "The rate of heat loss of a body is proportional to the difference in temperatures between the body and its surroundings." However, by definition, the validity of Newton's law of cooling requires that the rate of heat loss from convection be a linear function of ("proportional to") the temperature difference that drives heat transfer, and in convective cooling this is sometimes not the case.

In general, convection is not linearly dependent on temperature gradients, and in some cases is strongly nonlinear. In these cases, Newton's law does not apply.

Radiation

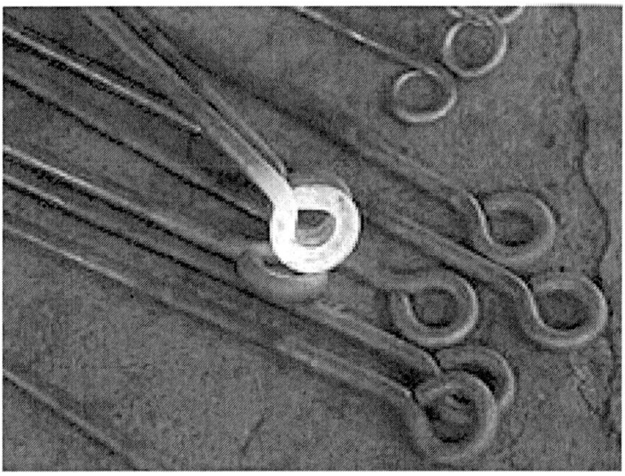

Figure 3: A red-hot iron object, transferring heat to the surrounding environment primarily through thermal radiation.

Thermal radiation is energy emitted by matter as electromagnetic waves due to the pool of thermal energy that all matter possesses that has a temperature above absolute zero. Thermal radiation propagates without the presence of matter through the vacuum of space.

Thermal radiation is a direct result of the random movements of atoms and molecules in matter. Since these atoms and molecules are composed of charged particles (protons and electrons), their movement results in the emission of electromagnetic radiation, which carries energy away from the surface.

Unlike conductive and convective forms of heat transfer, thermal radiation can be concentrated in a small spot by using reflecting mirrors, which is exploited in concentrating solar power generation. For example, the sunlight reflected from mirrors heats the PS10 solar power tower and during the day it can heat water to 285 °C (545 °F).

Mass Transfer

In mass transfer, energy—including thermal energy—is moved by the physical transfer of a hot or cold object from one place to another. This can be as simple as placing hot water in a bottle and heating a bed, or the movement of an iceberg in changing ocean currents. A practical example is thermal hydraulics.

Convection vs. Conduction

In a body of fluid that is heated from underneath its container, conduction and convection can be considered to compete for dominance. If heat conduction is too great, fluid moving down by convection is heated by conduction so fast that its downward movement will be stopped due to its buoyancy, while fluid moving up by convection is cooled by conduction so fast that its driving buoyancy will diminish. On the other hand, if heat conduction is very low, a large temperature gradient may be formed and convection might be very strong.

The Rayleigh number (*Ra*) is a measure determining the result of this competition.

$$Ra = \frac{g\Delta\rho L^3}{\mu\alpha} = \frac{g\beta\Delta T L^3}{\nu\alpha}$$

where

- g is acceleration due to gravity
- ρ is the density with $\Delta\rho$ being the density difference between the lower and upper ends
- μ is the dynamic viscosity
- α is the Thermal diffusivity
- β is the volume thermal expansivity (sometimes denoted α elsewhere)

- T is the temperature and
- ν is the kinematic viscosity.

The Rayleigh number can be understood as the ratio between the rate of heat transfer by convection to the rate of heat transfer by conduction; or, equivalently, the ratio between the corresponding timescales, up to a numerical factor. This can be seen as follows, where all calculations are up to numerical factors depending on the geometry of the system.

The buoyancy force driving the convection is roughly $g\Delta\rho L^3$, so the corresponding pressure is roughly $g\Delta\rho L$. In steady state, this is canceled by the shear stress due to viscosity, and therefore roughly equals $\mu V / L = \mu / T_{conv}$, where V is the typical fluid velocity due to convection and T_{conv} the order of its timescale. The conduction timescale, on the other hand, is of the order of $T_{cond} = L^2 / \alpha$. Convection occurs when the Rayleigh number is above 1,000–2,000. For example, the Earth's mantle, exhibiting non-stable convection, has Rayleigh number of the order of 1,000, and T_{conv} as calculated above is around 100 million years.

Phase Changes

Transfer of heat through a phase transition in the medium—such as water-to-ice, water-to-steam, steam-to-water, or ice-to-water—involves significant energy and is exploited in many ways: steam engines, refrigerators, etc. For example, the Mason equation is an approximate analytical expression for the growth of a water droplet based on the effects of heat transport on evapouration and condensation.

Boiling

Heat transfer in boiling fluids is complex, but of considerable technical importance. It is characterised by an S-shaped curve relating heat flux to surface temperature difference.

At low driving temperatures, no boiling occurs and the heat transfer rate is controlled by the usual single-phase mechanisms. As the surface temperature is increased, local boiling occurs and vapour bubbles nucleate, grow into the surrounding cooler fluid, and collapse. This is *sub-cooled nucleate boiling*, and is a very efficient heat transfer mechanism. At high bubble generation rates, the bubbles begin to interfere and the heat flux no longer increases rapidly with surface temperature (this is the departure from nucleate boiling, or DNB). At higher temperatures still, a maximum in the heat flux is reached (the critical heat flux, or CHF). The regime of falling heat transfer that follows is not easy to study, but is believed to be characterised by alternate periods of nucleate and film boiling. Nucleate boiling slows the heat transfer due to gas bubbles on the heater's surface; as mentioned, gas-phase thermal conductivity is much lower than liquid-phase thermal conductivity, so the outcome is a kind of "gas thermal barrier". At higher temperatures still, the hydrodynamically-quieter regime of film boiling is reached. Heat fluxes across the stable vapour layers are low, but rise slowly with temperature. Any contact between fluid and the surface that may be seen probably leads to the extremely rapid nucleation of a fresh vapour layer ("spontaneous nucleation").

Condensation

Condensation occurs when a vapour is cooled and changes its phase to a liquid. Condensation heat transfer, like boiling, is of great significance in industry. During condensation, the latent heat of vapourisation must be released. The amount of the heat is the same as that absorbed during vapourisation at the same fluid pressure.

There are several types of condensation:

- Homogeneous condensation, as during a formation of fog.
- Condensation in direct contact with subcooled liquid.
- Condensation on direct contact with a cooling wall of a heat exchanger: This is the most common mode used in industry:
    - Filmwise condensation is when a liquid film is formed on the subcooled surface, and usually occurs when the liquid wets the surface.
    - Dropwise condensation is when liquid drops are formed on the subcooled surface, and usually occurs when the liquid does not wet the surface.

Dropwise condensation is difficult to sustain reliably; therefore, industrial equipment is normally designed to operate in filmwise condensation mode.

Heat Equation

The heat equation is an important partial differential equation that describes the distribution of heat (or variation in temperature) in a given region over time. In some cases, exact solutions of the equation are available; in other cases the equation must be solved numerically using computational methods. For example, simplified climate models may use Newtonian cooling, instead of a full (and computationally expensive) radiation code, to maintain atmospheric temperatures.

Lumped System Analysis

System analysis by the lumped capacitance model is a common approximation in transient conduction that may be used whenever heat conduction within an object is much faster than heat conduction across the boundary of the object.

This is a method of approximation that reduces one aspect of the transient conduction system—that within the object—to an equivalent steady state system. That is, the method assumes that the temperature within the object is completely uniform, although its value may be changing in time.

In this method, the ratio of the conductive heat resistance within the object to the convective heat transfer resistance across the object's boundary, known as the *Biot number*, is calculated.

For small Biot numbers, the approximation of *spatially uniform temperature within the object* can be used: it can be presumed that heat transferred into the object has time to uniformly distribute itself, due to the lower resistance to doing so, as compared with the resistance to heat entering the object. Lumped system analysis often reduces the complexity of the equations to one first-order linear differential equation, in which case heating and cooling are described by a simple exponential solution, often referred to as Newton's law of cooling.

Applications and Techniques

Heat transfer has broad application to the functioning of numerous devices and systems. Heat-transfer principles may be used to preserve, increase, or decrease temperature in a wide variety of circumstances.

Insulation and Radiant Barriers

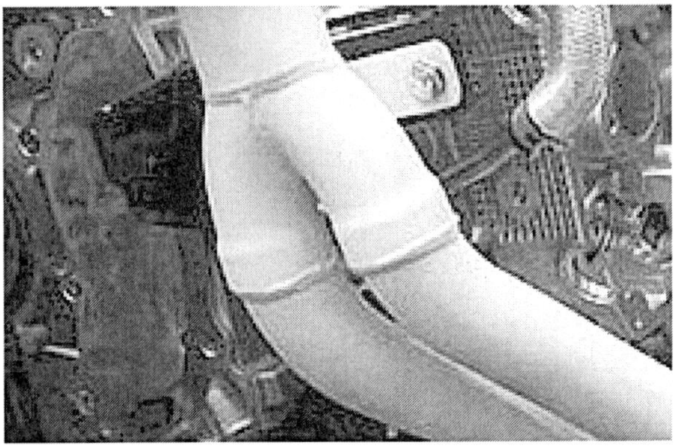

Figure 4: Car exhausts usually require some form of heat barrier, especially high performance exhausts where a ceramic coating is often applied.

Figure 5: Heat exposure as part of a fire test for firestop products.

Thermal insulators are materials specifically designed to reduce the flow of heat by limiting conduction, convection, or both. Radiant barriers are materials that reflect radiation, and therefore reduce the flow of heat from radiation sources. Good insulators are not necessarily good radiant barriers, and vice versa. Metal, for instance, is an excellent reflector and a poor insulator.

The effectiveness of an insulator is indicated by its R-value, or resistance value. The R-value of a material is the inverse of the conduction coefficient (*k*) multiplied by the thickness (*d*) of the insulator. In most of the world, R-values are measured in SI units: square-metre kelvins per watt ($m^2 \cdot K/W$). In the United States, R-values are customarily given in units of British thermal units per hour per square-foot degrees Fahrenheit ($Btu/h \cdot ft^2 \cdot °F$).

$$R = \frac{d}{k}$$

$$C = \frac{Q}{m\Delta T}$$

Rigid fibreglass, a common insulation material, has an R-value of four per inch, while poured concrete, a poor insulator, has an R-value of 0.08 per inch.

The tog is a measure of thermal resistance, commonly used in the textile industry, and often seen quoted on, for example, duvets and carpet underlay.

The effectiveness of a radiant barrier is indicated by its reflectivity, which is the fraction of radiation reflected. A material with a high reflectivity (at a given wavelength) has a low emissivity (at that same wavelength), and vice versa. At any specific wavelength, reflectivity

= 1 - emissivity. An ideal radiant barrier would have a reflectivity of 1, and would therefore reflect 100 percent of incoming radiation. Vacuum flasks, or Dewars, are silvered to approach this ideal. In the vacuum of space, satellites use multi-layer insulation, which consists of many layers of aluminised (shiny) Mylar to greatly reduce radiation heat transfer and control satellite temperature.

Critical Insulation Thickness

Low thermal conductivity (k) materials reduce heat fluxes. The smaller the k value, the larger the corresponding thermal resistance (R) value. Thermal conductivity is measured in watts-per-metre per kelvin ($W \cdot m^{-1} \cdot K^{-1}$), represented as k. As the thickness of insulating material increases, the thermal resistance—or R-value—also increases.

However, adding layers of insulation has the potential of increasing the surface area, and hence the thermal convection area.

For example, as thicker insulation is added to a cylindrical pipe, the outer radius of the pipe-and-insulation system increases, and therefore surface area increases. The point where the added resistance of increasing insulation thickness becomes overshadowed by the effect of increased surface area is called the critical insulation thickness. In simple cylindrical pipes, this is calculated as a radius:

$$R_{critical} = \frac{k}{h}$$

Heat Exchangers

A heat exchanger is a tool built for efficient heat transfer from one fluid to another, whether the fluids are separated by a solid wall so that they never mix, or the fluids are in direct contact. Heat exchangers are widely used in refrigeration, air conditioning, space heating, power generation, and chemical processing.

One common example of a heat exchanger is a car's radiator, in which the hot coolant fluid is cooled by the flow of air over the radiator's surface.

Common types of heat exchanger flows include parallel flow, counter flow, and cross flow. In parallel flow, both fluids move in the same direction while transferring heat; in counter flow, the fluids move in opposite directions; and in cross flow, the fluids move at right angles to each other. Common constructions for heat exchanger include shell and tube, double pipe, extruded finned pipe, spiral fin pipe, u-tube, and stacked plate.

When engineers calculate the theoretical heat transfer in a heat exchanger, they must contend with the fact that the driving temperature difference between the two fluids varies with position. To account for this in simple systems, the log mean temperature difference (LMTD) is often used as an "average" temperature. In more complex systems, direct knowledge of the LMTD is not available, and the number of transfer units (NTU) method can be used instead.

Heat Dissipation

A heat sink is a component that transfers heat generated within a solid material to a fluid medium, such as air or a liquid. Examples of heat sinks are the heat exchangers used in refrigeration and air conditioning systems, and the radiator in a car (which is also a heat exchanger). Heat sinks also help to cool electronic and optoelectronic devices such as CPUs, higher-power lasers, and light-emitting diodes (LEDs). A heat sink uses its extended surfaces to increase the surface area in contact with the cooling fluid.

Buildings

In cold climates, houses with their heating systems form dissipative systems. In spite of efforts to insulate houses to reduce heat losses via their exteriors, considerable heat is lost, which can make their interiors uncomfortably cool or cold. For the comfort of the inhabitants, the interiors must be maintained out of thermal equilibrium with the external surroundings. In effect, these domestic residences are oases of warmth in a sea of cold, and the thermal gradient between the inside and outside is often quite steep. This can lead to problems such as condensation and uncomfortable air currents, which—if left unaddressed—can cause cosmetic or structural damage to the property. Such issues can be prevented by use of insulation techniques for reducing heat loss.

Thermal transmittance is the rate of transfer of heat through a structure divided by the difference in temperature across the structure. It is expressed in watts per square metre per kelvin, or W/m^2K. Well-insulated parts of a building have a low thermal transmittance, whereas poorly-insulated parts of a building have a high thermal transmittance.

A thermostat is a device capable of starting the heating system when the house's interior falls below a set temperature, and of stopping that same system when another (higher) set temperature has been achieved. Thus, the thermostat controls the flow of energy into the house, that energy eventually being dissipated to the exterior.

Thermal Energy Storage

Thermal energy storage refers to technologies that store energy in a thermal reservoir for later use. They can be employed to balance energy demand between daytime and nighttime. The thermal reservoir may be maintained at a temperature above (hotter) or below (colder) than that of the ambient environment. Applications include later use in space heating, domestic or process hot water, or to generate electricity. Most practical active solar heating systems have storage for a few hours to a day's worth of heat collected.

Evapourative Cooling

Evapourative cooling is a physical phenomenon in which evapouration of a liquid, typically into surrounding air, cools an object or a liquid in contact with it. Latent heat describes the amount of heat that is needed to evapourate the liquid; this heat comes from the liquid itself and the surrounding gas and surfaces. The greater the difference between the two temperatures, the greater the evapourative cooling effect. When the temperatures are the same, no net evapouration of water in air occurs; thus, there is no cooling effect. A simple example of natural evapourative cooling is perspiration, or sweat, which the body secretes in order to cool itself. An evapourative cooler is a device that cools air through the simple evapouration of water.

Radiative Cooling

Radiative cooling is the process by which a body loses heat by radiation. It is an important effect in the Earth's atmosphere. In the case of the Earth-atmosphere system, it refers to the process by which long-wave (infrared) radiation is emitted to balance the absorption of short-wave (visible) energy from the Sun. Convective transport of heat and evapourative transport of latent heat both remove heat from the surface and redistribute it in the atmosphere, making it available for radiative transport at higher altitudes.

Laser Cooling

Laser cooling refers to techniques in which atomic and molecular samples are cooled through the interaction with one or more laser light fields. The most common method of laser cooling is Doppler cooling. In Doppler cooling, the frequency of the laser light is tuned slightly below an electronic transition in the atom. Thus, the atoms would absorb more photons if they moved towards the light source, due to the Doppler effect. If an excited atom then emits a photon spontaneously, it will be accelerated. The result of the absorption and emission process is to reduce the speed of the atom. Eventually the mean velocity, and therefore the kinetic energy of the atoms, will be reduced. Since the temperature of an ensemble of atoms is a measure of the random internal kinetic energy, this is equivalent to cooling the atoms.

Sympathetic cooling is a process in which particles of one type cool particles of another type. Typically, atomic ions that can be directly laser-cooled are used to cool nearby ions or atoms. This technique allows cooling of ions and atoms that cannot be laser cooled directly.

Magnetic Cooling

Magnetic evapourative cooling is a technique for lowering the temperature of a group of atoms. The process confines atoms using a magnetic field. Over time, individual atoms will become much more energetic than the others due to random collisions, and will escape—removing energy from the system and reducing the temperature of the remaining group. This process is similar to the familiar process by which standing water becomes water vapour.

Heat Transfer in the Human Body

The principles of heat transfer in engineering systems can be applied to the human body in order to determine how the body transfers heat. Heat is produced in the body by the continuous metabolism of nutrients which provides energy for the systems of the body. The human body must maintain a consistent internal temperature in order to maintain healthy bodily functions. Therefore, excess heat must be dissipated from the body to keep it from overheating. When a person engages in elevated levels of physical activity, the body requires additional fuel which increases the metabolic rate and the rate of heat production. The body must then use additional methods to remove the additional heat produced in order to keep the internal temperature at a healthy level.

Heat transfer by convection is driven by the movement of fluids over the surface of the body. This convective fluid can be either a liquid or a gas. For heat transfer from the outer surface of the body, the convection mechanism is dependent on the surface area of the body, the velocity of the air, and the temperature gradient between the surface of the skin and the ambient air. The normal temperature of the body is approximately 37°C. Heat transfer occurs more readily when the temperature of the surroundings is significantly less than the normal body temperature. This concept explains why a person feels "cold" when not enough covering is worn when exposed to a cold environment. Clothing can be considered an insulator which provides thermal resistance to heat flow over the covered portion of the body. This thermal resistance causes the temperature on the surface of the clothing to be less than the temperature on the surface of the skin. This smaller temperature gradient between the surface temperature and the ambient temperature will cause a lower rate of heat transfer than if the skin were not covered.

In order to ensure that one portion of the body is not significantly hotter than another portion, heat must be distributed evenly through the bodily tissues. Blood flowing through blood vessels acts as a convective fluid and helps to prevent any buildup of excess heat inside the tissues of the body. This flow of blood through the vessels can be modelled as pipe flow in an engineering system. The heat carried by the blood is determined by the temperature of the surrounding tissue, the diametre of the blood vessel, the thickness of the fluid, velocity of the flow, and the heat transfer coefficient of the blood. The velocity, blood vessel diametre, and the fluid thickness can all be related with the Reynolds Number, a dimensionless number used in fluid mechanics to characterise the flow of fluids.

Latent heat loss, also known as evapourative heat loss, accounts for a large fraction of heat loss from the body. When the core temperature of the body increases, the body triggers sweat glands in the skin to bring additional moisture to the surface of the skin. The liquid is then transformed into vapour which removes heat from the surface of the body. The rate of evapouration heat loss is directly related to the vapour pressure at the skin surface and the amount of moisture present on the skin. Therefore, the maximum of heat transfer will occur when the skin is completely wet. The body continuously loses water by evapouration but the most significant amount of heat loss occurs during periods of increased physical activity.

Other

A heat pipe is a passive device constructed in such a way that it acts as though it has extremely high thermal conductivity. Heat pipes use latent heat and capillary action to move heat, and can carry many times as much heat as a similar-sized copper rod. Originally invented for use in satellites, they have applications in personal computers.

A thermocouple is a junction between two different metals that produces a voltage related to a temperature difference. Thermocouples are a widely used type of temperature sensor for measurement and control, and can also be used to convert heat into electric power.

A thermopile is an electronic device that converts thermal energy into electrical energy. It is composed of thermocouples. Thermopiles do not measure the absolute temperature, but generate an output voltage proportional to a temperature difference. Thermopiles are widely used, e.g., they are the key component of infrared thermometres, such as those used to measure body temperature via the ear.

A thermal diode or thermal rectifier is a device that preferentially passes heat in one direction: a "one-way valve" for heat.

Bibliography

Annand, W.J. and G.E. Roe: *Gas Flow in the Internal Combustion Engine*, Haessner Publishing, Inc., 1974.

Ansdale, R.F. : *The Wankel RC Engine: Design and Performance*, Iliffe Books, Ltd., London, 1968.

Ashby, E.: *Technology and the Academics*, Macmillan, London, 1959.

Ayres, Robert U. and Richard P. McKenna: *Alternatives to the Internal Combustion Engines*, Johns Hopkins University Press, 1972.

Badiru, A.: *Handbook of Industrial and Systems Engineering*, CRC Press, Delhi, 2005.

Basshuysen, R. and F. Schaefer: *Internal Combustion Engine Handbook.* Warrendale, PA: SAE, 2004.

Beizer, Boris: *Software Testing Techniques,* International Thomson Computer Press, Boston, MA., 2004.

Bennett, S.: *A History of Control Engineering: 1800-1930,* IEE Press, London, 1979.

Benson, R.S. and N.D. Whitehouse: *Internal Combustion Engines*, Pergamon Press, Inc. 1979.

Bigelow, J.: *Elements of Technology*, Boston Press, Boston, 1829.

Bird, John: *Electrical and Electronic Principles and Technology,* Newnes, UK, 2007.

Blackmore, D.R. and A. Thomas: *Fuel Economy of the Gasoline Engine*, John Wiley & Sons, 1977.

Borman, G.L. and K.W. Ragland: *Combustion Engineering*, WCB McGraw-Hill, 1998.

Brittain, A., and P. Morton: *Engraving on Precious Metals*, N.A.G. Press, London, 1958.

Bucciarelli, Louis L.: *Designing Engineers*, MIT Press, Cambridge, 1994.

Burns, G.; Glazer, A.M.: *Space Groups for Scientists and Engineers,* Academic Press, Boston, 1990.

Bye, G. C.: *Portland Cement Composition Production and Properties*, Pergamon Press, Oxford, 1983.

Cebon, David: *Materials: Engineering, Science, Processing and Design*, Butterworth-Heinemann, UK, 2007.

Chapman, K.: *The International Petrochemical Industry*. Oxford: Blackwell, 1991.

Checkel, M. D. and J. D. Dale: *Computerised Knock Detection from Engine Pressure Records,* Warrendale, PA: Society of Automotive Engineers, 1986.

China, T.: *Building Industry, Pride of the Pacific,* Honolulu, Hawaii, 1991.

Craig, Bruce D. and David S. Anderson: *Handbook of Corrosion Data. Materials Park,* OH: ASM International, 1995.

Crandall, Richard: *Projects in Scientific Computation,* NY: Springer- Verlag, 1994.

Cummins, Lyle : *Internal Fire: The Internal Combustion Engine 1673 - 1900,* Society of Automotive Engineers, 1976.

Dasgupta, S.: *Technology and Creativity*, Oxford University Press, New York, 1996.

Davey, N.: *History of Building Materials*, Phoenix House, London, 1961.

David Cebon: *Materials: Engineering, Science, Processing and Design,* Butterworth-Heinemann, UK, 2007.

David Willoughby: *Plastic Piping Handbook*. McGraw-Hill Professional, Delhi, 2002.

Dimes, Francis B.: *Conservation of Building and Decorative Stone,* Heineman, Butterworth, 1990.

Dorf, Richard C.: *Electrical Engineering Handbook*. Boca Raton, FL: CRC Press, 1997.

Eastop, T. D. and A. McConkey: *Applied Th ermodynamics for Engineering Technologists,* Essex, UK: Longman Group, 1995.

Edward H. Smith: *Mechanical Engineer's Reference Book,* Butterworth-Heinemann, Oxford, 1994.

Fabrycky, W.: *Systems Engineering and Analysis*, Prentice-Hall, NY, 2005.

Fayette Taylor, C. and Edward S. Taylor: *The Internal Combustion Engine*, International Textbook Company, 1961.

Ferguson, C.R. : *Internal Combustion Engines—Applied Thermosciences*, John Wiley & Sons, 1986.

Fitchen, J.: *Building Construction Before Mechanization,* MIT Press, London, 1988.

Francois Cardarelli: *Scientific Unit Conversion: A Practical Guide to Metrication*, Springer, M.J. Shields. London, 1997.

Gibson, Dan: *The Nabataeans, Builders of Petra,* CanBooks, Amman, Jordan, 2002.

Goodger, E.M. : *Hydrocarbon Fuels; Production, Properties and Performance of Liquids and Gases*, Macmillan, London, 1975.

Gossweiler, C. R., W. Sailer, and C. Cater. *Sensors and Amplifi ers for Combustion Analysis.* Winterthur: Kistler Instruments AG, 2005.

Greene, A. B. and G. G. Lucas: *Th e Testing of Internal Combustion Engines.* Bucks, UK: English University Press, 1969.

Grimwade, Mark: *Introduction to Precious Metals*, Newnes Technical Books-Butterworth, London, 1985.

Hamlin, T.: *Architecture Through the Ages*, G. P. Putnam's Sons, New York, 1953.

Haworth, R. : *Combustion Noise from High Speed Direction Injection Diesel Engines,* Warrendale, PA: Society of Automotive Engineers, 1985.

Heisler, H.: *Advanced Engine Technology.* Warrendale, PA: SAE, 1995.

Heywood, J. B.: *Internal Combustion Engine Fundamentals.* Columbus: McGraw-Hill, 1988.

Hollands, J.: *Engineering Psychology and Human Performance*, Prentice Hall, Delhi, 1999.

Horlock, J.H. and D.E. Winterbone, *The Thermodynamics and Gas Dynamics of Internal Combustion Engines*, Clarendon Press, Oxford, 1986.

James Brown: *Advanced Machining Technology Handbook,* McGraw-Hill, New York, 1998.

Jones, R.T., Cohen, D.: *High Speed Wing Theory*, Princeton University Press, 1960

Kuratle, R. H. : *The State of the Art Combustion Engine Pressure Instrumentation.* Winterthur: Kistler Instruments A.G, 1995.

Kutz, M.: *Mechanical Engineers' Handbook*, John Wiley and Sons, New York, 1986.

Lichty, L.C. : *Combustion Engine Processes*, McGraw-Hill Book Company, 1967.

Lifschitz, E. M.: *Fluid Mechanics*, Pergamon Press, North Holland, 1987.

Makartchouk, A. : *Diesel Engine Engineering: Thermodynamics, Dynamics, Design, and Control, Marcel Dekker*, Inc., New York, Basel, 2002.

Mark Pinsky: *Introduction to Fourier Analysis and Wavelets,* Brooks-Cole, 2001.

Martyr, A. J. and M. A. Plint. *Engine Testing,* Oxford: Elsevier, 2007.

Mugerwa, A. K. : *Fuel Cell Systems.* New York: Plenum Press, 1993.

Nunney, M.J. : *The Automotive Engine*, Newnes-Butterworths, London, 1974.

Nuti, M. : *Emissions from Two-Stroke Engines*, SAE, 1998.

Obert, F. : *Internal Combustion Engines and Air Pollution*, Intext Educational Publishers, 1973

Okill, J. *Autographic Indicators for Internal Combustion Engines.* London: Edward Arnold & Company, 1938.

Ott, James: *Jets: Airliners of the Golden Age.* Pyramid Media Group, 1990.

Parsons, William Barclay: *Engineers and Engineering in the Renaissance,* Williams & Wilkins, Baltimore, 1939.

Patterson, D.J. and N.A. Henein: *Emission from Combustion Engines and their Control,* Ann Arbor Science Publishers, Inc., 1972.

Poonia, M P and S K Mehla: *Biodiesel : Performance Evaluation of Internal Combustion Engines,* Aavishkar Pub, Delhi, 2008.

Pulkrabek, W.W. : *Engineering Fundamentals of the Internal Combustion Engine,* Prentice-Hall, Inc., 1997.

Ramalingam, K.K. : *Internal Combustion Engines,* Scitech Pub, Delhi, 2009

Randolph, A. L. : *Methods of Processing Cylinder Pressure Transducer Signals to Maximize Accuracy,* Warrendale, PA: Society of Automotive Engineers, 1990.

Rathi, Rakesh : *Petroleum Refining Processes,* SBS Pub, Delhi, 2007.

Rogowski, A.R. : *Elements of Internal Combustion Engines,* McGraw-Hill, 1953.

Sher, E. : *Handbook of Air Pollution from Internal Combustion Engines: Pollutant Formation and Control,* Academic Press, 1998.

Singh K & Soler A: *Mechanical Design of Heat Exchangers,* Arcturus, London, 1984

Sitkei, G. : *Heat Transfer and Thermal Loading in Internal Combustion Engines,* Akademiai Kaido:Budapest, 1974.

Springer, G.S. and D.J. Patterson: *Engine Emissions: Pollutant Formation and Measurement.* Plenum Press, New York, London, 1973.

Stone, R.: *Introduction to Internal Combustion Engine Fundamentals,* Hampshire, UK: Palgrave, 1999.

Thomson, W. : *Fundamentals of Automotive Engine Balance,* Mechanical Engineering Publications, Ltd., London, 1978.

Treager, I.E.: *Aircraft Gas Turbine Engine Technology.* . McGraw-Hill Book Co. Glencoe Div., 1995.

Waddell, J. J.: *Construction Materials Ready Reference Manual,* McGrawHill Book Co., New York, 1985.

Watson, N. and M.S. Janota, Turbocharging the Internal Combustion Engine, John Wiley & Sons, New York, 1982.

Winterbone, D.E. and R.J. Pearson, *Theory of Engine Manifold Design,* SAE, 2000.

Young, C.D. : *Measurement of Diesel Combustion Noise,* Bloomfi eld Hills, MI: Autotech, 1985.

Zhao, H. and N. Ladommatos: *Engine Combustion Instrumentation and Diagnostics.* Warrendale, PA: SAE, 2001.

Index

E

F

G

H

I

J

L

M

O

P

Q

R

❍❍❍